CONCORDIA UNIVERSITY
DD231.L8T7　　　　　　C001 V
LUDENDORFF BOSTON

3 4211 000023110

LUDENDORFF

LUDENDORFF the soldier was forced into the position of Dictator — this, according to Karl Tschuppik, was the main cause of Germany's defeat. Again and again the Chancellor or the Reichstag handed decisions over to this expert in military affairs, regardless of the fact that the soldier never understands the necessity of subordinating action to political gain. Bismarck, in 1870, fought tooth and nail with the generals, but this time there was no Bismarck to fight Ludendorff.

Although the war is described from the German point of view, the author shows the Allied conduct of affairs and how Allied statesmen managed to control generals. Of highest importance as a document and as an analysis, this book also makes excellent reading. The whole conduct of the war is summarized in its pages, rising to a climax with the operations of 1918 in France, 'the greatest battle in history.'

LUDENDORFF
THE TRAGEDY OF A MILITARY MIND

BY

KARL TSCHUPPIK

TRANSLATED BY

W. H. JOHNSTON, B.A.

BOSTON AND NEW YORK
HOUGHTON MIFFLIN COMPANY
1932

*The German original, "Ludendorff, die Tragödie des Fachmanns"
was published in Vienna in 1930*

All rights reserved

PRINTED IN GREAT BRITAIN
BY R. & R. CLARK, LIMITED, EDINBURGH

CONTENTS

CHAPTER I

BEGINNINGS 1
Early Career—Schlieffen and Cannae—The Masurian Lakes and Tannenberg.

CHAPTER II

THE RISE TO SUPREMACY 19
The Marne—The Problem of Schlieffen—Falkenhayn's Appointment—Operations on the Eastern Front—The Triumvirate of Kaiser, Statesman, and General—Falkenhayn's Fall and Ludendorff's Rise.

CHAPTER III

DICTATORSHIP 58
First Visit to Western Front—Roumanian Campaign—Politician v. General—Submarine Warfare.

CHAPTER IV

DICTATORSHIP (*continued*) 88
Submarine Warfare—Tanks—First Retreats—Failure of Submarines—Fall of Bethmann Hollweg.

CHAPTER V

LUDENDORFF AND THE POLITICIANS 111
Failure of the Politicians—The Danger of the Expert—Michaelis and the Peace Resolution—The Campaign of 1917—The Pope's Peace Appeal—First Success of the Tanks.

CHAPTER VI

LUDENDORFF AND THE POLITICIANS (*continued*) . . . 137
  The Collapse of Russia—Missed Opportunities—Failure of the Reichstag—Lack of a Peace Programme—Ludendorff and Hoffmann—Threat of Resignation—Diminished Rôle of the Kaiser—The Peace of Brest-Litovsk.

CHAPTER VII

THE BEGINNING OF THE END 165
  Modern Warfare—Plans for 1918—The Battle of March 1918.

CHAPTER VIII

GROWING DIFFICULTIES 192
  Retrospect—Criticism—Growing Difficulties—Battle of April 1918—The Cost of the Offensive.

CHAPTER IX

THE LAST STRAW 211
  Battle of May 1918—Kühlmann's Fall—German War Aims, July 1918.

CHAPTER X

DEFEAT 224
  Third Battle of the Marne—The Black 8th of August—Failure of the Politicians—The Future of Belgium.

CHAPTER XI

COLLAPSE 240
  Lack of Plans for a Defensive—Bulgaria Surrenders—No Hope Left—Constitutional Troubles—Note to President Wilson—Ludendorff a Last Ditcher—Ludendorff's Resignation.

CHAPTER XII

THE END 269

INDEX 279

MAPS—The Russian Front . . . . *facing page* 19
  The Western Front . . . . ,, 165

CHAPTER I

BEGINNINGS

Early Career—Schlieffen and Cannae—The Masurian Lakes and Tannenberg

It is the early morning of August 7th, 1914; a German officer strikes with the pommel of his sword upon the gates of Liége. Major-General Ludendorff is alone except for his aide-de-camp. The chauffeur of the Belgian car who brought the two officers to the spot watches them with amazement—surely the man does not hope to capture a citadel with a bare sword? The General and his aide-de-camp had expected to find the citadel in the hands of the advance guard of their brigade; but not a single German soldier is in sight. The garrison of the citadel answer the knocking and open the gates. The two officers continue on their audacious course and, undeterred by the sight of some hundreds of Belgian soldiers, penetrate into the courtyard. The defenders are paralysed by surprise; soon after, a German regiment is heard approaching, and the garrison surrenders to General Ludendorff.

The important fortress of Liége, whose function it was to guard the approaches of the valley of the Meuse, had fallen without one of its outer line of forts having been touched. Twelve powerful forts, hitherto regarded as impregnable, guarded the fortress; not one of them was in the hands of the Germans. General von Emmich with a force of six weak brigades at his disposal, succeeded in carrying only one of these between the forts and into the citadel. To this brigade Ludendorff had attached himself, and when the commanding officer fell in action Ludendorff took over the command. It was in this capacity that he went through his first battle and saw his first man killed. On a later occasion he noted in his diary: "I shall never forget the sound made by a bullet striking a human body".

The night of the 6th to the 7th of August passed in unendurable suspense. Alone within the circle of forts and without news of the other units, the brigade might at any moment be cut off and surrounded. In spite of this danger, Emmich gave orders in the morning to occupy the town. The advance guard, which had been ordered to occupy the citadel, lost its way, and when Ludendorff appeared before the walls he was alone with his aide-de-camp. In later days he referred to the advance upon Liége and the capture of this city as the "most treasured memory" of his life.

It was an accident which had given to Ludendorff this position in the forefront of the battle. He admits himself that he had been intended to take part in the advance on Liége in the capacity of a casual spectator and without holding any executive post. His real place was with the Staff of the Second Army, where he had been attached to General von Lauenstein, the Chief of General Staff, as Deputy-Chief of Staff. This appointment corresponded roughly to Ludendorff's age and past career. For some years he had served with his infantry regiment at Wesel, Wilhelmshaven, Kiel, Frankfurt-an-der-Oder, and Thorn; later he had been stationed as General Staff Officer at Glogau and Posen, and from 1904 to 1913 he had worked in the offices of the General Staff at the War Office, where he had spent eight years in the Mobilisation Department, finishing this part of his service as head of the department. At this time plans for an increase in the army were being worked out, and in his capacity as head of his department Ludendorff had been working since 1912 with all the energy and tenacity at his disposal for the incorporation of three new Army Corps. It was this self-imposed advocacy of the cause of the army which led to his relegation from the General Staff: he was given the command of a brigade of infantry stationed at Düsseldorf. In April 1914, at the age of forty-nine, he obtained a brigade at Strassburg. On the outbreak of war he was attached to the Second Army.

BEGINNINGS

During the advance on Liége, Ludendorff was acting as liaison officer. It was his function to maintain contact between General von Bülow, commanding the Second Army, and the Storm Troops of General Emmich. After the capture of the fortress it was intended that he should return to the Army Staff. Had he done so, he would have advanced with Bülow through Belgium and Northern France to the Marne. As it was, Liége was for Ludendorff what Toulon was for the Young Officer of Artillery.

In 1793, while the French Revolutionary Government was carrying through its policy of general conscription, every officer who showed signs of boldness and talent might hope to rise to the rank of general, while generals who met with misfortune were not saved from the Tribunals of a Republic by their personal courage. Stendhal remarks that this apparently absurd system, which offered a suitable target for the sarcasms of all the Legitimists in Europe, provided France with all her greatest commanders. Once promotion had become a matter of routine, the system produced nothing but mediocrity—generals like Macdonald, Oudinot, Dupont, and Marmont, whose armies between 1808 and 1814 achieved for Napoleon nothing but defeats.

It is worth asking whether Stendhal's criticism of Revolutionary France can be applied to the armies of Imperial Germany, and whether age and seniority were of greater importance than boldness and talent. The name of Ludendorff had been familiar in the High Commands and among all the officers of the General Staff, but since 1913 he had fallen into oblivion. It was Liége and the Order *pour le mérite* which brought Ludendorff back from obscurity and into favour. As the dangers of a war on two fronts began to emerge, men began to look towards Ludendorff, and on the 22nd of August he was instructed to take up his command on the Eastern front.

"A difficult task is being entrusted to you, one more difficult perhaps than the capture of Liége", Moltke, the Chief of General Staff, wrote to Ludendorff. "I know of no other person whom I trust as implicitly as yourself. Perhaps you may succeed in saving the position in the East. . . . Your energy is such that you may still succeed in averting the worst." And

General von Stein, the Quartermaster General, accompanied Moltke's appeal with the words: "Your place is on the Eastern front. The safety of the country demands it." This appeal in the hour of trouble must have sounded to Ludendorff like the voice of justice. "I know of no other person"—could it be that fate was beginning to make good the faults of the past? Ludendorff calls back to his mind the privations of his life: "My parents were far from wealthy, and their faithful work was not rewarded by riches on earth. We lived in a very small way. . . . My father and mother were fully occupied with anxiety for their six children. . . . When I was a young officer I had a hard struggle of it to make my way." He then touches upon a particularly painful spot. For a long period he had been destined, in the event of war, to be the head of the directorate of military operations. His transfer to Düsseldorf had put an end to this plan. It was Ludendorff's successor in the Great General Staff, Colonel von Tappen, who became the assistant of Moltke.

Ludendorff's appointment as Chief of Staff of the Eighth Army in Eastern Prussia was one to satisfy the loftiest ambition. "It was an inspiring thought", Ludendorff writes, "that at this crucial moment I was serving my country at a decisive post." The German force in the East consisted of four Army Corps, one Reserve Division, and one Cavalry Division; it was opposed by two Russian armies, each of which was superior to the German army of the East. The unexpected arrival of General Samsonov's Warsaw army, which marched at night and remained under cover of the forests during the day, compelled the German Commander-in-Chief, General von Prittwitz, to break off the operations against the Vilna army commanded by General Rennenkampf and to withdraw behind the Vistula. A little later Lieutenant-Colonel Hoffmann, a member of his Staff, induced him to change his view of the situation, and von Prittwitz himself gave the first orders for an advance against the Warsaw army; but it was now too late, and both he and his Chief of Staff, Count Waldersee, were recalled. On the 23rd of August, at four o'clock in the morning, the new Commander-in-Chief on the Eastern front, von Hindenburg, joined Ludendorff in his

BEGINNINGS

train at Hanover. This was the first meeting between Hindenburg and his Chief of Staff.

The two men whom the train was carrying from Hanover to Marienburg are not surrounded by that glamour which a kind of traditional superstition loves to attach to military leaders. Ludendorff himself would have declined to enter into abstract considerations upon the nature of generalship. He was devoid of that mysticism which demands superhuman qualities and the protection of higher powers for the captain who rides the whirlwind and directs the storm. Ludendorff knew no uncertainty; his confidence in himself and his abilities rested upon a sure foundation of acquired knowledge and inborn personal qualities. In every detail he was the pupil of Alfred von Schlieffen, the last great master who moulded the form of the Prussian General Staff. von Schlieffen was too clear a thinker and too fair-minded a critic of the German strategy of the 19th century not to know that a general is not made by training or by appointment. The shepherd David, destined to be the conqueror of the Philistines, was anointed by Samuel to command the King's army, and Hannibal was dedicated as a child to the career of arms upon the altar of Baal. Caesar attributed his good fortune at Dyrrhachium to superhuman powers, and Cromwell felt himself to be the chosen instrument of the Lord. Schlieffen disposes of these fancies in a brief sentence worthy of Frederick: "If the incipient general relies upon his divine call, his genius, or the protection and support of higher powers, his hopes of victory are scanty". Ludendorff's sentiments were the same. The general's task is the destruction of the enemy, even if the latter's forces are stronger and his position and intentions unknown. A happy gift of improvisation does not suffice for this; the unknown can be mastered only by the skilful application of the stored-up wisdom of the Prussian General Staff.

Clausewitz has laid it down as a general proposition that the differences between the nations cannot be formulated in an epigram and can be found only in the sum total of their intellectual and material relations to each other. This general proposition is true in particular of the differences between the methods of warfare. Every nation in arms carries beneath its uniform its acquired characteristics and its past history, and a general, although he may imagine himself to be a free agent, is in fact the servant of a tradition no less strong than the walls that encircle a fortress. Only a man of genius succeeds at times in breaking through these bonds.

The world for Ludendorff was of a Spartan simplicity: he knew no uncertainty and no problems. The son of an obscure Pomeranian family never doubted that the Prussian world which he first saw as a child and learned to know as a cadet and officer was the best of all possible worlds. He had not the critical eyes which, while not blind to the defects of creation, are necessary also in order to discern the essential from the irrelevant and to pierce through the trappings to the man. From early youth he had been trained to receive an order and to obey, and his mind became immune alike to speculation and to doubt. Any flaws and faults in the well-ordered picture of the Prussian State were interpreted by Ludendorff from an ethical point of view: they were no more than the effects of the infirmity of individuals. Surely the rise of Prussia was its own justification; surely the Prussian General Staff wielded the most admirable instrument of war, and surely the principles of Schlieffen possessed infallibility. Every staff officer of Ludendorff's generation had absorbed the principles of Schlieffen, and their validity was universal, like that of the laws of mathematics. If Gneisenau was the founder of the Prussian General Staff, and the elder Moltke had inspired it with new life, then Schlieffen's function had been to perfect the tradition of Moltke. The General Staff becomes a complicated technical office whose function is the conduct of war.

If we discard technical language, Schlieffen's guiding idea can be easily explained. It is based on the fact, proved by experience, that it is easier to overcome the enemy by attacking him in flank or rear than by meeting him face to face. Boys when they play at soldiers or Red Indians arrive at this advantageous method without the help of a military college. In these games, that group will be in the best position which arranges its forces so that one part faces and attacks the enemy while the two other parts advance on his right and left and threaten him on either flank and in the rear. Old and simple as is this rule, it has been practised successfully only by captains of genius.

It was Hans Delbrück, the great military historian, who reconstructed the battle of Cannae, and thus enabled Schlieffen to make this classical example of the battle of so-called double outflanking the ideal of all campaigns in general. On the 2nd of August 216 B.C., Hannibal, with 50,000 men, had taken up a position near the village of Cannae on the Aufidus, in the Apulian plain. He was opposed by 69,000 men under Terentius Varro. Faced by a superior enemy and with the sea in his rear, Hannibal was in a more than unfavourable position. The Romans had three lines of battle, the *hastati*, the *principes*, and the *triarii*; this was a classification by age, and the three classes corresponded approximately to the troops with the colours, the reservists, and the *Landwehr* of a modern army. Where this arrangement was followed, only the first line actually fought, carrying its reserve formations with it "like a snail carrying its house upon its back". These reserve formations waited under arms behind the first line until their turn should come, and it was only then that the second and, in case of need, the third maniple went into action. Thus the full forces of a Roman army were never engaged simultaneously, and only a fraction was in action at any given time. At Cannae, Hannibal, faced by this rigid and clumsy mass, whose reserves waited passively until the fighting formations had been decimated, divided his army into three mobile groups which could be brought into action simultaneously. His forces were weaker than those of the Romans; his front line, consisting of infantry, was drawn up in shallow formation with cavalry on either wing; behind them were drawn up his best troops, the heavily armed Carthaginians. In the first shock of battle Hannibal's infantry was thrust back by the sheer weight of the Roman assault. This movement, however, was brought to a stop as soon as the Carthaginian attacks on the two Roman flanks and the cavalry charge delivered upon their rear made themselves felt. The Roman square, exposed to an attack from every side, was eventually crushed in, with the loss of 48,000 men, while Hannibal's losses only amounted to 6000.

This battle, fought over 2000 years ago, became Count Schlieffen's ideal. He proceeded to investigate military history, and particularly the campaigns of Frederick the Great and Napoleon, in an attempt to see how closely their methods approximated to the ideal battle of annihilation. His rigorous criticism of Moltke's two campaigns, that against Austria and that against France, had the one aim of demonstrating that Moltke's plans invariably aimed at inflicting a Cannae upon his enemy: where the ideal of a battle of annihilation was not attained, the fault did not lie with the Chief of the General Staff but with the refractory Army Commanders. The whole trend of thought of the General Staff and the will of budding generals came to be trained with one aim in view, and with one aim alone—Cannae.

During the journey to Marienburg, Ludendorff had leisure to examine the prospects of the impending battle. His position was less favourable than Hannibal's. The enemy was more than twice as strong as his forces, a disadvantage attaching inevitably to a war waged on two fronts; for while the army of the West was attacking the enemy with every available man, the army of the East had to remain on the defensive until the troops from the West should be set free. The position, moreover, was far worse than had been anticipated. The Russian mobilisation had been unexpectedly rapid, and Moltke's letter to Ludendorff did not exaggerate. "Perhaps you will be in time to avert the worst." The prospect was dark indeed.

The earliest plans for a war on two fronts had been worked out under Schlieffen, who had foreseen a withdrawal behind the Vistula in case of necessity. The rapid advance of Samsonov's army endangered this line of defence, since it was possible that he might reach the Vistula before the Germans. A comparison of forces, and every other military consideration, made a successful defence appear improbable and a German victory impossible.

Marienburg, the site of Army Headquarters, was reached

by Ludendorff on the 23rd August. He knew that in an all but desperate position he had only one chance. The two Russian armies, commanded respectively by Rennenkampf and Samsonov, were separated by the chain of the Masurian Lakes. The distances, however, were inconsiderable: Rennenkampf was within 30 miles of the retiring German corps and within 55 miles of Samsonov. The German army of the East was thus in a position analogous of that of Benedek in 1866. If Ludendorff followed academic routine and attacked one of the two Russian armies with the whole of his forces, the other army might be attacking his own rear within two days. Thus everything turned on the question whether the two German corps facing Rennenkampf could be withdrawn and made available for a general attack against Samsonov. Ludendorff had already drawn every man who could be spared from the various garrisons, and if he succeeded in gathering all his forces against Samsonov before Rennenkampf began to move, there was a chance that Samsonov might be defeated. Even this, however, was only the first and less difficult half of Ludendorff's task: after the battle with Samsonov it would be necessary to turn his tired troops against the fresh forces of Rennenkampf.

Military history contains no example of a position as desperate as Ludendorff's. According to Stendhal's simple definition, the art of Napoleon consisted in arranging matters in such a way that on the actual field of battle his men outnumbered the enemy by two to one. His principle is identical with that of a pair of footpads who attack a single person, and thus outnumber him by two to one, despite the fact that a dozen policemen are on their beat a hundred yards away. Once the victim has been robbed, it is little use if the police arrive three minutes after the event. At Rovereto and Bassano and in numerous engagements during the Tyrolese campaign a thousand Frenchmen uniformly sufficed to defeat three thousand Austrians; on the actual field of battle, however, the French equally uniformly outnumbered the Austrians by two to one. Ludendorff, however, was so placed that even the boldest manœuvre could at best enable him to attack the enemy with equal numbers, while the other half of the enemy remained intact and continued to threaten his rear. Once Rennenkampf set himself in motion everything would be lost.

Ludendorff's critics, including Hans Delbrück and the French critic Pierrefeu, maintain that it was a gambler's risk to hazard an engagement in such a perilous position. Ludendorff accepted the risk. He relied upon the irresolution of the Russians, and he was favoured by a piece of carelessness on their part: both the Russians armies made a practice of sending their wireless messages *en clair*, and the Germans succeeded in picking them up. The single Cavalry Division belonging to the German army of the East was left to obstruct Rennenkampf, while the remaining forces advanced against Samsonov. Without this dangerous manœuvre it would have been impossible to defeat the two Russian armies; but Ludendorff had no alternative.

A clear outline of the impending battle had by now formed in his mind, and his instructions and orders were all directed towards a single end. But the most brilliant conception could be upset by an unforeseen accident, and as soon as the first shot was fired everything became subject to chance. The battle proper began at dawn on the 27th of August, Ludendorff's aim being to encircle the army of General Samsonov. His own troops had just completed long and exhausting marches; a number of units had sustained losses in the engagements with Rennenkampf, and the roads were encumbered with fugitives who made the regular movement of troops difficult. Yet in spite of these difficulties Ludendorff aimed at the maximum possible result—the encirclement and destruction of the enemy.

Samsonov's five corps were marching in a north-westerly direction: the attack was delivered upon them in the extensive forest region south of Allenstein, a district of some thirty miles square. It was made from two directions—from the west and from the north. But although Ludendorff's general aim was to inflict a Cannae upon the enemy, he did not go so far as to aim at a complete encirclement; his flanking movement went too far east to embrace the most southerly Russian corps. General Hoffmann describes this decision of Ludendorff's as the decisive point of the whole battle of Tannenberg. On the first day the troops at the centre could only advance with difficulty between Hohenstein and Grossgardinen, while Mackensen and Bülow were successful in the north. On the second day Mackensen continued to advance in the rear of the enemy, and in the afternoon the centre began to advance. By now victory seemed certain, although it was still uncertain whether it would take the shape of a Cannae. Three-quarters of the ring encircling the Russians had by now assumed definite form.

The morning of the 29th of August brought bad news. A flying officer reported that the Russian corps which, as mentioned above, had remained outside the German encircling movement was now advancing. General Artamanov had rightly resolved to attack the German right wing in an endeavour to relieve the pressure upon the Russian army. The Germans, however, succeeded in checking Artamanov, while at the same time continuing their encircling movement northward. Meanwhile in the centre of the German front, east of Hohenstein, a number of German columns were converging upon a common objective; between them were masses of Russian prisoners of war. The four Russian corps were driven together from the four quarters of the compass and all attempts to break the ring in the south remained fruitless. On the evening of the 29th of August the image of Cannae had been completed. Near Willenburg, General Samsonov was found among the dead—seeing disaster inevitable, he had shot himself. Ninety thousand Russians were taken prisoner.

Ludendorff has recorded that the victory of Tannenberg did not give him undiluted satisfaction; always at the back of his mind there remained the fear of Rennenkampf's army. "I thought of General von Schlieffen and thanked the Master."

Before the brilliant victory of Tannenberg, Ludendorff had been informed that General Headquarters had resolved to attach three Army Corps of the Western army to the army of the East. The events have been recorded by General Hoffmann in his *Memoirs*. Colonel Tappen, the head of Operations Section at General Headquarters, asked for Ludendorff on the telephone. Ludendorff said to Hoffmann: "You might take the second earpiece, so that you can hear what Tappen is saying and how I reply". Tappen's message was to the effect that three Army Corps and one Cavalry Division of the Western army were to be transferred to the East. Tappen further asked where they were to be sent. Ludendorff explained his wishes, but at the same time pointed out that he was not in need of reinforcements and that, if there were any difficulties in the West, he was prepared to do without them. To this Tappen replied that the forces could be spared in the West. Next day Colonel Tappen rang up again, and once more Hoffmann listened-in on the field telephone. Tappen declared that he could spare only two corps and one Cavalry Division, and that the third corps promised was wanted in the West. Ludendorff answered that the battle against Rennenkampf had already been engaged and that the troops would be too late in any case: if they were needed in the West, they need not trouble to worry about the East. In his *Memoirs*, Ludendorff tells us that he "never asked for reinforcements at all". In his opinion, the resolve to withdraw troops from the West was premature, a fact which "we in the East, unfortunately, were not in a position fully to appreciate". Ludendorff's version has never been contradicted. It is important, because the two corps which were drawn from the Western front may be held in part responsible for the disaster of the Marne.

While the drama of the Marne was beginning in the West, Ludendorff struck at Rennenkampf in the East. The battle of the Masurian Lakes was bold in conception and execution. Once again Ludendorff was faced by greatly superior forces; 192 German battalions were opposed by 384 Russian battalions. In these circumstances, the ideal—to outflank the enemy on both flanks—was unattainable. The geographical position and the respective forces of the opponents alike made a manœuvre of this nature impracticable. Ludendorff's plan of battle was as follows. Four corps, their left wing resting upon the river Pregel, advanced upon the front of the enemy, while two corps and two Cavalry Divisions, constituting the right wing of his forces, advanced in a north-easterly direction in an attempt to outflank the enemy. The decisive blow would be struck by the flank corps of this wing, which was

BEGINNINGS

commanded by General François. The pressure exerted by François upon the flank and rear of the enemy paralysed the resistance of the Russian front, despite the fact that strong defences had been constructed there. Rennenkampf resolved to break off the engagement and to retreat. The bulk of his army succeeded in crossing the Niemen to safety, but even so 45,000 Russians were captured. Yet Ludendorff himself admits that the results of the battle were not as striking as those of Tannenberg. "What was lacking was the pressure upon the enemy's rear." The question has been debated by military critics whether Ludendorff's outflanking wing would have been more successful if it had been stronger. This does not appear likely; in such an event, Rennenkampf probably would merely have been induced to evade the final decision somewhat earlier than he did in fact.

The victories of Tannenberg and of the Masurian Lakes were the foundation of Ludendorff's fame. Few battles of the Great War have won so much praise and have had so many legends attached to them; few have been equally misrepresented by interested parties. The popular fiction of what might be called the patriotic help rendered by the morasses was destroyed by Ludendorff himself. "The story that thousands of Russians were driven into the morasses and perished is a myth. For miles around there is no morass in existence."

Matters are different in the controversy about Ludendorff's share in the crucial decision at Tannenberg. The controversy is due to General von François, the Commander of the German 1st Army Corps, and has recently been revived by Mr. Churchill's analysis of the battle. Churchill attaches very great importance to François's share in the victory of Tannenberg, and says that the credit for this victory will for all time belong to General François, who led his corps with a rare combination of prudence and audacity, skilfully evading Ludendorff's orders and winning an overwhelming victory against the latter's instructions. Churchill is undoubtedly one of the greatest experts on the military history of the Great War, and has made a careful study of the battle of Tannenberg. At the same time, his judgment in assigning all the praise to the prudent Corps Commander and allowing no credit to the Commander-in-Chief, cannot be described as objective: it is merely the expression of a subjective opinion—and military history is full of these, like every other type of history.

At the battle of Tannenberg the task of General von François was to attack Samsonov's army of the Narev from the west and to interfere with its deployment. The forces at his disposal consisted of his own Army Corps and General Mühlmann's detachment, and the task was assigned to him because his forces constituted the southernmost group of the Eighth Army. His corps had begun to detrain at Deutsch-Eylau on the 23rd of August. On this day certain parts of the 20th Army Corps, commanded by General Scholtz, were already in action at Hohenstein, some distance east of Deutsch-Eylau. François's orders were to deliver the decisive attack on the 28th of August. The following is General Hoffmann's account: "Here again a difference of opinion with General François arose. The General desired to postpone the attack by one day because part of his troops had not yet arrived, and further he desired to make his flanking attack more sweeping and in the direction of Mlava. In the opinion of the Army Commander, time was of supreme importance; every day that we lost might enable Rennenkampf to get under way. Further, if the German Eighth Army was extended so far as to outflank the left wing of Samsonov's army at Mlava, such an operation would have tended to scatter the German forces, which were all too weak already. Accordingly, the Army Commander ordered the break through to be made at Usdau—which, in my opinion, was the decisive point of the whole battle of Tannenberg."

Ludendorff gives a similar account in different words: "The army of the Narev was advancing in echelon on the left. . . . The southernmost echelon on the 26th of August had reached the region of Waplitz. Further on the left and in the rear, and in a somewhat more westerly position, the 1st Russian Army Corps was moving by way of Mlava and Soldau. . . . The problem was to interfere with this movement from the West. There was a considerable temptation in doing so to make a southerly flanking movement embracing Soldau, and with it the whole of the 1st Russian Army Corps. If this were done, the defeat of the army of the Narev ... might become overwhelming. The forces at our disposal, however, were inadequate, and accordingly I suggested to General von Hindenburg to attack in the direction of Usdau, using the 1st Army Corps, commanded by François and stationed at Deutsch-Eylau and Montowo, and the 20th Army Corps, stationed at Gilgenburg. The idea was to thrust back the 1st Russian Army Corps in a southerly direction from Soldau. This done, the 1st Army Corps was to penetrate in the direction of Neidenburg, and to join the 17th Army Corps and the 1st Reserve Corps in surrounding the main body of the army of the Narev. It was essential to restrict our aims if we were to hope for success. ... The attack of the 1st and the 20th Army Corps had to be postponed to the 27th of August. I would have been pleased to see it undertaken earlier, but the inadequate railway system of Eastern Prussia made it impossible for the 1st Corps to be ready in time. The General commanding the 1st Army Corps, von François, rightly insisted on having his whole forces on the spot before attacking."

The following, on the other hand, is von François's version: "On the 25th of August, Hindenburg and Ludendorff called upon me at Battle Headquarters. I was instructed to make a frontal attack upon the hills of Usdau on the morning of the 26th. I objected that by that time the bulk of my artillery and ammunition columns would still be on the way, and that consequently my corps would not be ready for action. To this Ludendorff remarked: 'In that case you must attack with the infantry alone'. There followed a somewhat acrimonious discussion, which Ludendorff concluded by saying: 'The corps must attack'. At the same time he referred the matter to Hindenburg by saying: 'The final decision, however, rests with the Commander-in-Chief'. The latter made no remark and both left my Battle Headquarters. At 8.30 P.M. I received orders in writing, signed by Hindenburg, instructing me to attack the lines of Usdau on the 26th not later than 10 A.M. Once more I voiced my objection, but once more was overruled. To attack Usdau without artillery and ammunition would have been a tactical blunder which might have led to the annihilation of my corps, for the Russians were holding the position with three divisions. . . . On the 26th the advance of our forces was delayed, and I was thus enabled to postpone the attack of Usdau until the morning of the 27th. My corps was now ready for action and the attack was successful."

General von François gives very exact details with regard to the difference of opinion about the time for delivering the attack. Ludendorff had good reason for urging the importance of a rapid attack. We saw that Hoffmann remarks that "every day that we lost might enable Rennenkampf to get under way. . . ." Certain destruction was impending during the whole of the battle. Eventually François had his way, and Ludendorff admitted that he was right. The "tactical blunder" never arose. But the essential point, the difference of opinion on the strategical question, is passed over in silence by François. The latter wished to carry out a wide flanking attack by way of Mlava, while Ludendorff, desiring a less extensive operation, ordered a break through to be attempted at Usdau. The Army Commander's aim was essentially correct; it is this that Hoffmann describes as the decisive point of the battle of Tannenberg; but the only words which François finds to describe Ludendorff's plan are that a frontal attack was ordered—words which contain an implied criticism. The second instance adduced by François to prove that the credit for Tannenberg belongs to him and not to Ludendorff relates to the order given after the break through at Usdau. "On the morning of the 28th of August", General François records, "I received (at Soldau) urgent orders from Hindenburg immediately to despatch a division to support the 20th Corps (General Scholtz). The 41st Division, which formed part of his corps, was supposed to have been defeated and in retreat. I despatched the 2nd Division and ordered the 1st Division, which had captured Soldau, to advance on Willenburg by way of Neidenburg in order to cut off the retreat of the Russian army of the Narev. This operation was being carried on, when a new order arrived instructing me to take up the pursuit of the Russians in the direction of Lahna. Lahna is six miles north of Neidenburg, and an advance in that direction would have led me into wooded and hilly country where artillery would have been useless and the troops could only have advanced very slowly. The gravest objection, however, was that such a movement would have afforded the Russians a golden opportunity of escaping towards the South. Consequently I did not change the orders I had already given, and gave instructions to have this non-compliance with orders reported with my reasons to Army Headquarters. In consequence of these measures my troops captured 60,000 unwounded prisoners and 231 guns."

General von François's account tells us nothing that we did not know already. Ludendorff and Hoffmann themselves, in their accounts of the battle, emphasise the fact that the decisive task had been assigned by Ludendorff to the 1st Army Corps. It was against the wishes of von François that Ludendorff enforced the idea of a limited Cannae and the plan of leaving the 1st Russian Corps outside his flanking movement, confining himself to driving it south outside this encircling ring: this piece of prudent moderation was Ludendorff's. But military history relates dozens of examples of a conflict of wishes between subordinate commanders and their Commanders-in-Chief. Old Moltke was hardly ever able to realise his ideas according to plan. It will remain to the credit of von François that he obeyed and that in the execution of his orders he excelled Army Headquarters in tactical foresight. In conducting the pursuit of the Russians via Neidenburg instead of the more northerly Lahna he was doing no more than his obvious duty—he had better opportunities of knowing the terrain and the actual position of the battle than Army Headquarters, fixed as it was behind an advancing front. von François would have cause to bring a charge against Ludendorff before the eyes of critics and of history if Ludendorff had opposed or disregarded the better information of the Corps Commander. This, however, was not the case.

Accordingly, Churchill is not correct in assigning the greater part of the praise to François, as he does when he says that the latter prudently and deliberately refrained from complying with Ludendorff's orders. "Refrained from complying" is not applicable here, since it implies a hostile rivalry which, although unfortunately François was to some extent guilty of it, cannot be assumed to be a desirable relation between a Commander-in-Chief and his Corps Commander. The fact is that François substituted better orders for those given by Ludendorff. Students of German military history will be aware that it is due to the tradition of Moltke that Corps Commanders were left a maximum amount of liberty within the limits of their general task; it was left to them to give the orders demanded by the general idea of the action in conformity with their local knowledge. Especially was this true in the case of a pursuit, where the only person capable of properly appreciating the position is the subordinate commander.

The mention of the elder Moltke recalls the fact that General von François had certain qualities in common with his predecessor, General von Steinmetz. Steinmetz was equally capable and opinionated; brilliant alike in attacking the enemy and in carrying on a campaign against his superior commanders. In 1866 Steinmetz had opened the approaches to Bohemia to the army of the Crown Prince, yet in 1870, after the victorious battle of Spichern, Moltke caused him to be sent home. It was a tradition of the German army to set a high value on the independence of military judgment; even higher importance was, however, attached to discipline. In the Austrian army things were different; here independent non-compliance with orders was rewarded with the highest decoration, the Maria Theresa Order, provided that such action met with success. History has taught us that this laudable policy had poor results in practice.

CHAPTER II

THE RISE TO SUPREMACY

The Marne—The Problem of Schlieffen—Falkenhayn's Appointment—Operations on the Eastern Front—The Triumvirate of Kaiser, Statesman, and General—Falkenhayn's Fall and Ludendorff's Rise

THE war had now been in progress for six weeks, and it was impossible to foresee the dimensions it might attain. The drama of the Marne could not as yet be properly appreciated. One person alone had broken down mentally and physically, Count Moltke, the Commander-in-Chief of the German forces. He had been a broken man from the moment when, in the last hour, the Kaiser had expressed his criticism of the advance through Belgium. The English resolve to make war had become apparent and, terrified, the Kaiser had asked the Chief of General Staff whether the German plans of advance could be altered. The only answer possible for Moltke was in the negative; no alternative plan was available. The Kaiser said: "Your uncle would have given a different answer".

The Kaiser's criticism touched a fatal flaw in the German plans. Four years later similar remarks were made by critics, historians, politicians, and arm-chair strategists. The elder Moltke, on whose name the Kaiser thus called in the hour of danger, had on two occasions, in 1870 and in 1880, considered the possibility of a war on two fronts. In the second plan the intention was to attack the Russians with the main body and to take up a defensive attitude against France with the smaller force. Moltke had been induced to adopt this plan by reasons of a military, political, and psychological nature. Under his successor, Count Waldersee, a number of different plans were considered. Recollections of 1812 and the difficulty of a Russian winter made it appear doubtful whether Moltke's plan was worth retaining, and Waldersee elaborated two plans according to the season of the year. If war broke out in summer, the main attack was to take place in the East. If it broke out in winter, the main attack was to be made in the West.

It was Schlieffen who finally abandoned Moltke's plan. From the accounts given by Schlieffen's disciples it is clear that he was convinced that the assumptions which guided Moltke in drawing up his plan no longer were valid. Since Moltke's time the military power of France had grown very considerably, and the French army, full of the spirit of the offensive, was supposed to have become the more dangerous enemy. Hence the rules of strategy demanded that the attack must be delivered against France. The most thoughtful critic of German strategy, Count Hans von Hentig, comments on this that it is an injustice to Moltke to assume that this skilful strategist was counting on any lack of an aggressive spirit in the French army. Moltke knew that France would attack in any case, for the aim of France in the event of war would be to recapture Alsace-Lorraine, and this aim implied the will to take the offensive. But Moltke also took it for granted that the Russians would not content themselves with remaining on the defensive. If Moltke resolved to remain on the defensive in the West, and to take up a position east of the Rhine with the Main covering his left flank for the decisive battle, this decision was due to the consideration that a rapid and victorious peace with Russia would break the warlike spirit of France.

In any case, this belated return to Moltke was not made until certain initial errors had been committed. In 1914 the German armies marched according to Schlieffen's amended plan. It may be worth asking what is the psychological origin of the attitude inspiring the plans and principles of Schlieffen. It would be erroneous to assume that it was the "self-confidence of a rapidly advancing nation" which gave birth to the idea of the strategical annihilation of the enemy; for a consciousness of power and even an aggressive spirit does not

THE RISE TO SUPREMACY

make it impossible that a nation might be controlled by a general whose chief qualities were prudence and deliberation. The history of Germany affords examples of such a state of affairs. In any case, the obvious fact that Schlieffen drew his principles from the study of military history is of greater importance than the search for the psychological motives of the military idea underlying a military system. Schlieffen reached his ultimate idea after prolonged study of the facts—his absolute belief in the idea of a Cannae inspired him with the plan of a gigantic outflanking battle in France. This belief was so powerful as to overcome political objections. Without unflanking there could be no victory, and there could be no outflanking without a converging advance; the resolve to march through Belgium is necessarily implied in the idea of a Cannae. It followed logically that in Schlieffen's plan the right wing of the army of the West was made as powerful as possible. This wing, consisting of 35 Army Corps and 8 Cavalry Divisions, was to advance through Belgium into Northern France, and executing a vast wheel towards the south, to outflank the enemy. The left wing, to which Schlieffen assigned four and a half Army Corps and 3 Cavalry Divisions, would be faced by the main French attack and would be compelled to withdraw. The hinge around which the two German army groups revolved was in the region of Diedenhofen and Metz. Schlieffen was not afraid that the French might succeed in reaching the Rhine. The further east the French penetrated, the greater would be the danger for them, and while they were advancing upon the Rhine the decisive battle would be fought near Paris. It was not until then that the German left wing, reinforced by troops drawn from the right, would assume the offensive and finally grip the French between two converging forces.

This plan was modified by Schlieffen's successor, the younger Moltke. The left wing of the army of the West received 8 Army Corps, and the idea of a retreat before the French army of the Rhine in Alsace-Lorraine was banished. This plan aimed at "checking or defeating" the enemy on the left wing. Such fundamental alterations completely changed the outline of the decisive battle in France. According to

Schlieffen, the outflanking movement was entrusted to one wing, and for this purpose the right wing was made as powerful as possible; according to Moltke, both the wings were to co-operate in seizing the enemy in a grip. We know that the forces available for the German armies in the West were inadequate for this plan. According to Schlieffen, the right wing of the German army could not have been too powerful; as it was, two corps belonging to this wing were held before Antwerp and one before Maubeuge, while two further corps were transferred to the East. In spite of this weakening of the right, no reinforcements were drawn from the left to strengthen it, and at the decisive moment the right wing, whose outflanking movement was to have embraced Paris, was too short. Moltke fully appreciated the danger threatening from Paris, and tried to adapt himself to the altered circumstances. He attempted to turn the two armies of the extreme right wing against Paris and to deliver the decisive attack between Toul and Epinal while slowly advancing at the centre. These plans were rendered impossible by events, and the surprise attack delivered by Maunoury against the extreme German left wing decided the issue of the battle of the Marne, an issue very different from Schlieffen's anticipations.

Military critics have seen in the battle of the Marne the beginning of the end: "the germs of the collapse of Germany". It was at the Marne that the plan devised by the General Staff for dealing with the difficulties of a war on two fronts was brought to nothing. Every defeat leads to a painful process of self-accusation in which an attempt is made to trace back the fault to one individual cause: all attempts to explain the drama of the Marne come back to the German plan of deployment. Why was Schlieffen's plan altered?

The *Memoirs* of Erzberger tell us that General Moltke sadly confessed in January 1915 that his plans had been wrong, and that it would have been right to keep to his uncle's idea, who intended to throw the bulk of the German army against Russia and to remain on the defensive in the West. But this view is inspired by the fact of defeat, and it is opposed by Ludendorff, who claims that even the bulk of the German army could not have forced a rapid decision in the East, and that, if such an attempt had been made, a large part, if not the whole, of the most important German industrial regions would have been seized by the enemy. Germany would have been paralysed, and it was the plans of Schlieffen and not of the elder Moltke that should have been followed.

At this point Hans Delbrück's criticism begins. Delbrück does not reject the elder Moltke's plans; according to him they were a workable alternative to Schlieffen's plan, even in the altered circumstances, and he goes on to say that either of the plans might have proved successful if followed to its logical conclusion. What was disastrous was the modification of Schlieffen's plan. Its essence was the strength of the right wing, and when this idea was abandoned the plan "was deprived of its soul". Delbrück proceeds to accuse Ludendorff of having "diluted" Schlieffen's plan, since it was Ludendorff who was head of the Operations Section from 1908 to 1913, during which period the new plan was worked out. To this charge Ludendorff has replied. In his opinion, as opposed to that of Delbrück and other theorists, it would have been possible to strengthen the left wing of the German army without making the right wing less powerful than Schlieffen had intended. Ludendorff thinks that there was reason for anxiety about the left wing, since an advance of the enemy between Metz and Strassburg might have broken the German lines of communication before the victory had been decided on the right. The younger Moltke further feared the ruin of the industrial region north of Saarbrücken, and he considered this danger so serious that he felt it his duty to take steps to prevent it. He saw no danger in doing so, because an attack by the German left wing offered the chance of defeating a considerable part of the French army at the very outset of war, and of preventing them from being used in stemming the German advance. Ludendorff further says that Schlieffen's plan was not watered down by these arrangements, pointing out that since the pre-war increase in the number of reserve divisions, Moltke possessed forces sufficient to admit of a strengthening of the left wing without reducing the number of divisions destined for the right. Moltke's arrangements prevented a critical position arising on the left wing and, what in Ludendorff's opinion was of crucial importance, made possible a decisive victory. The reason why the plan failed is not to be looked for in the modifications which Schlieffen's plans underwent, but in the failure of the High Command.

It was the German plan of advance that decided the success of the German army in a war on two fronts, and given the old ways of thought and the old military hierarchy this was a problem that defied solution. Ludendorff accepts it as an inevitable fact, "that, in view of the neglect of the German army, the decision on the Western front would have to be sought with inferior forces", adding that better training and a more competent Staff would have afforded no more than a partial compensation. It was the military historian, Hugo Schulz, who pointed out many years before the war that a militia system was not only a political but a military necessity for Germany. At a time when Clausewitz's idea of an "absolute war" dominated every General Staff, and when modern armaments had reached their actual pitch of perfection, a militia system was the only possible method of organising an army, since it was the only system which made it possible to place the entire forces of the nation in the field at the very outset. A German army of this kind would have had the power of carrying through the battle at Paris according to Schlieffen's plan.

But Ludendorff is on the wrong track when he accuses the rulers of Germany of having neglected their colonial forces instead of following the French and English in embodying coloured troops in their army. The cause of the disaster must be looked for elsewhere. Whereas in France every available man was under arms, in Germany half of the population capable of bearing arms was neither conscripted nor trained. The four million men who might have assured victory on the Marne remained at home, and it was not until after the decisive battle had been fought that they were given an inadequate training and consumed in a vain process of attrition during four years of trench warfare. To utilise the entire population capable of bearing arms and to give every man a full training was a financial impossibility for Germany. A militia system would have constituted a way out, and would have given Germany the security needed in a war on two fronts.

Ludendorff further replies to the critics who make him responsible for the modification of Schlieffen's plan that his period of responsibility at the Great General Staff was limited. It was not Ludendorff, but the Deputy-Chief of the General Staff, von Stein (at a later period Minister for War), who advised Moltke on questions of strategy. And indeed it is probable that in Ludendorff's position even a Scharnhorst would have been unable to break through the limitations imposed by the Prussian General Staff. At that time it would have required a revolutionary to grasp the superiority of the militia system, proved as it had been by the experiences of the Boer War and on the battlefields of Manchuria, and the General Staff was not the place for revolutionaries. Ludendorff's picture of the army, and of its mentality and lines of thought, shows the impossibility of so far-reaching a reform of the German forces. The General Staff was organised on the principle of a strict division of labour, and it was the practice to look upon the opinion of a superior as an inviolable command. In such surroundings fruitful and critical thought was impossible. Even under the old military system the new plan for the West was an operation which might and which did meet with criticism; but criticism was strictly limited, and this was not altered by the fact that valid reasons might be found in favour of an alteration of Schlieffen's plan.

To-day we know more. The reasons why Schlieffen's plan was modified were not of a military but of a political and psychological order. The Kaiser and his entourage could not reconcile themselves to the idea of placing Alsace-Lorraine, in whole or in part, in danger of even a temporary French invasion. The uncompromising strategists of the post-war period describe such an attitude as a crime and treat it as a "characteristic of the era of William II.", to be contrasted with the ideal of a purely unsentimental warfare as practised in the days of William I. and the elder Moltke.

Such a judgment cannot be accepted by a critic whose knowledge of Prussian history is drawn from other sources than patriotic text-books: the period of William I. provides us, in the treatment of Moltke's plans in 1866, with an exact analogy for the alteration of Schlieffen's plans under William II. Unless a general is in the fortunate position of Frederick the Great, Charles XII., or of Napoleon, and is ruler as well as Commander-in-Chief, he will always be in danger of being compelled to amend his plans. The elder Moltke's plans for the campaign of 1866 had been dictated by facts which imposed themselves. The points of detrainment followed from the position of the railways; it was intended to cross the frontier regions before the enemy had time to deploy his forces, whence it followed that Moltke had to enter Bohemia with a number of distinct forces which were to be united after the frontier had been passed. His ideal was to keep the distances to be covered as short as possible, and consequently he gave orders for the Second Army (commanded by Crown Prince Friedrich) to march into Bohemia via Landeshut. This was the first plan, which might have made it possible to force a Cannae before the end of June, and the only motives which led to its alteration were of a sentimental nature. The Crown Prince and a number of other persons besought the King not to leave Silesia defenceless before an Austrian invasion, and it was in vain that Moltke tried to convince him that Silesia was best protected by an offensive in Bohemia, and very badly, if at all, by a defensive on the spot. It had been represented to the King that it was his duty, as guardian of his people, to provide for the direct and visible defence of Silesia, and King William did not possess the necessary strength of mind to meet such an appeal with the reasons advanced by Moltke. The plan had to be altered; the Second Army took up a position behind Neisse, and, since two Army Corps were insufficient to hold the new front, a third Army Corps was despatched to the East. Schlieffen comments on this: "If anything could endanger Silesia, it was the measures taken for its protection". At that time neither Moltke nor the office of a Chief of General Staff possessed the authority to overcome the King's resolve. Moltke himself did not yet enjoy a great reputation, while other personal influences were all the more powerful. His carefully elaborated plans were reversed, and it was left to the author

THE RISE TO SUPREMACY

to build up a new structure from the fragments. He was victorious in spite of this disadvantage: the Austrian army was without leaders, the Austrian rifle was out-of-date, and the Austrian infantry tactics were based on the same idea as those of the Romans at Cannae, that of a number of columns distributed in depth. Even without Moltke's plan the Prussians possessed an overwhelming advantage, and they were spared the necessity of conducting an enquiry into the alteration of Moltke's plan. Schlieffen alone was not afraid to speak the truth.

Comparisons cannot make good a mistake once committed, but they can help to render apparent the traditional psychological compulsion leading to the commission of mistakes. The methods which might have endangered success in 1866 became fatal in 1914, and adapting the words of Schlieffen, we may say that if anything dealt a death-blow to Alsace-Lorraine it was the well-intentioned but ill-executed aim of saving its soil from a French invasion.

Colonel Ludendorff was not in a position to counteract such an aim. The wish of the King was equivalent to an order, and the order was executed by the Chief of General Staff. All that Ludendorff had to do was to draw up the plans according to the general outline given him. We can understand why he attached such importance to his demand for three new Army Corps. But his insistence was ill rewarded. Ludendorff left the General Staff and was attached to the infantry.

The critics who try to make Ludendorff in part responsible for the alteration of Schlieffen's plan are oblivious of the facts. They are assuming that a subordinate officer has powers of contradiction which have existed in no army, and least of all in the Prussian army. Napoleon had carried the campaign of 1796 to a victorious conclusion, he had won Milan and Lombardy and had acquired fame at Lodi, when the Directorate gave him the absurd instructions to divide his forces, an order "worthy rather of an enemy of the country". Napoleon replied with a long despatch to the Directorate, appealing to the common sense of this responsible body: "If by this division of forces you weaken the power of France, if . . . you destroy the unity of the guiding military thought, then

I must say, to my grief, that you will lose the fairest opportunity of imposing your will upon the enemy". But these urgent words were written in vain, and Napoleon was compelled to threaten that he would lay down the command.

To Ludendorff, an obscure colonel on the General Staff, it was not open to use the same language as the victor of Lodi. To have handed in his resignation would have been a mere gesture, and the only result of such a sacrifice would have been a brief notice in Army Orders. Again, a gesture like that of Yorck after the battle of Laon in March 1814 was unthinkable in the army of William II. On that occasion Yorck—the whole object of whose attack had been to effect a decisive battle—saw himself frustrated by the timidity of Gneisenau, and, deeply incensed, left Blücher's headquarters. On that occasion Blücher's tact and sense of solidarity was stronger than his sense of discipline. Blücher, himself a sick man, appealed to Yorck in a generous letter. "Old Comrade, do not leave the army. . . . I am very ill. . . . Come back, my good old Yorck." Marshal Foch, on whom this scene had made a deep impression, remarks in a certain passage that the "real and profound cause of Napoleon's defeat" is to be found in the humanity of Blücher, which allowed common sense to overcome discipline.

There is an old and rather fine saying that the German soldier carries Kant's moral doctrines with him in his valise. It can be paraphrased more correctly by saying that every army goes to war under conditions which are fixed beforehand, since it is controlled by the "sum total of the intellectual and material qualities of the nation". Among these qualities is the appreciation of intellectual moments in the army, by the capacity of the King to appreciate men of talent and to open himself to their arguments. During the struggle for the hegemony of Germany, Prussia had a ruler who possessed an eye for talent. The success of the triumvirate composed of King, Statesman, and General obscured the fact that its success depended solely upon the King.

The most important problem by the middle of September 1914 was to find a new Chief of Staff, a post which the Kaiser eventually filled by appointing the Prussian

THE RISE TO SUPREMACY

Minister for War, Erich von Falkenhayn. On the 14th of September, General Ludendorff had received orders to leave Hindenburg's Staff. His new field was to be smaller than that which he had had in the army of Hindenburg: he was to proceed to Breslau as Chief of Staff in the new army, which consisted only of two Army Corps. General Headquarters had been unable to form a comprehensive view of the position in the East. Ludendorff had overcome two Russian armies; but meanwhile the main forces of the Russians had overcome the Austro-Hungarian army. The magnitude of this defeat was not appreciated by General Headquarters, and the army of Upper Silesia, which was to support the Austrians, was meant to receive no more than two Army Corps. Ludendorff objected to this plan and suggested that the main body of Hindenburg's forces should be transferred to Upper Silesia and Posen. He himself could better appreciate the danger threatening from the Russian forces; but even Ludendorff did not get a view of the real facts until his journey to Austrian Headquarters took him through Galicia. Arrived at Neu Sandec, where the Austrian Headquarters were situated, he realised how ignorant the two allies were of each other. This was yet another instance of inadequate preparation. The two Chiefs of Staff, Moltke and Conrad, had arranged in writing and orally the steps to be taken in case of war, but these arrangements were quite vague and far from amounting to a regular plan of campaign. The two armies operated independently. Of the Austrian Chief of Staff, Ludendorff formed a very high opinion, but his army he did not consider "forcible enough". In Austria this remark, and later criticisms made by Ludendorff, caused offence, despite the fact that when Conrad was making his apology before the Imperial Military Cabinet he was compelled to speak of his troops in much harsher terms: "Only those who have experienced it can know what it is to work with an instrument that breaks in one's hand". It must further be admitted that the Austrians, like the Germans, were at a considerable disadvantage, 520 Austrian being opposed by 750 Russian battalions.

Meanwhile, Ludendorff's suggestion to give the Austrian

army more powerful support in its dangerous position had been sanctioned by Falkenhayn. The major part of Hindenburg's forces proceeded to Upper Silesia with their Commander and formed a new army—the Ninth. The rest (Eighth Army) remained for the protection of Eastern Prussia. Thus Hindenburg and Ludendorff were united again, and the Commander-in-Chief had the satisfaction of knowing that he was seconded by a man who would give intelligent execution to his plans. These plans and ambitions, however, were subject to limitation. The great question was whether the main attack was to be made in the East now that the offensive in the West had failed, and the answer to this question neither Ludendorff nor Hindenburg could influence. His victories in the East did not give Ludendorff sufficient authority to make himself heard in the supreme War Council, and Falkenhayn proceeded to prepare the Ypres offensive, which "will remain for all times the prototype of a battle entered upon hurriedly with inadequate Staff work and coming to nothing after enormous waste of valuable lives". Even within the framework of his new task Ludendorff was fettered more than he had been at Tannenberg and the Masurian Lakes. He was tied to the Austrian army, and the chief purpose of the impending battle was to extricate it from its cramped position between the Vistula and the Carpathians.

The five Army Corps at Ludendorff's disposal were not sufficient for a comprehensive movement against the army of the Grand Duke Nicolai Nicolaievitch. The Austrians were compelled to seek a decision south of the Vistula, to relieve Przemysl and to cross the San, while it was Ludendorff's task to draw a maximum number of Austrians from the Russian front. This could only be brought about by a rapid advance on the Vistula. Unless the Russians were engaged on the east bank of the Vistula before crossing it— and there was nothing to prevent them from doing so—the Germans would be too weak to offer a successful resistance. The Grand Duke divined the plan underlying the German advance and modified his own plans accordingly. Hitherto his forces had been united in the south and in the north—in

THE RISE TO SUPREMACY

Galicia and in Eastern Prussia; now he gathered more than thirty Army Corps between Ivangorod and the mouth of the San and advanced north along the Vistula as far as Warsaw and Novo Georgievsk. Having reached this position, he determined to outflank the forces commanded by Hindenburg and Ludendorff. He still had sufficient men to attack the Austrians on their front and to prevent them from moving. If the plan succeeded, a Russian victory was certain.

On the 12th of October the German troops were close to Warsaw. The immediate aim, to reach the Vistula, had been fulfilled; but the German forces were not strong enough to capture the two fortresses guarding the Vistula, Warsaw and Ivangorod, while at one point—near Kosjenitze—the Russians had already succeeded in crossing the river. The dangers of the position before Warsaw varied with the Austrian fortunes on the San. If the Austrians failed to advance or were forced to retreat, the line of the Vistula would become untenable.

On the 15th of October, Russian cavalry crossed the Vistula between Plotzk and Novo Georgievsk and advanced on the German rear. In this position Ludendorff wished to strengthen his rear by withdrawing three Army Corps from the line of the Vistula and putting Austrian troops in their place. Every moment was valuable, and in order to save time in deploying it was intended to draw the Austrian forces direct from Warsaw. But the Austrian Army Commander refused to comply with any request in this sense, and telegrams sent by Hindenburg to Falkenhayn and to the Kaiser, urging them to ask the Emperor Francis Joseph to give his sanction, were unavailing. Even this hour of danger failed to bring about a unity of command, and the utmost that Conrad could be induced to grant was that three German corps south of the Pilitza should be replaced by Austrian troops. The permission was given too late. The danger of being outflanked in the north had become imminent, and the pressure of the Russians on Novo Georgievsk and Warsaw was growing hourly. On the 17th of October Ludendorff was compelled to withdraw Mackensen's army from Warsaw, on the 21st of October the Austrians were defeated at Ivangorod and the line of the Vistula became untenable, and on the 27th of October a general retreat commenced. The retreat had been anticipated and was carried through according to plan, with the Russians slowly pursuing. The whole campaign had lasted four weeks.

Critics have found a good deal of fault with these operations. Hans von Hentig considers that they were carried through prematurely and without imagination. "The Austrians were not yet ready to fight, and the five German Army Corps who supported them were much too weak." He further urges that Ludendorff did not fulfil his function of covering the flank of the Austrian advance, and that his own advance on Warsaw was far too ambitious. "In military matters, where the lives of thousands may be at stake at any hour, it is impossible to feel regard for the commander who asks for the impossible. After exactly a month of costly operations the Austro-German forces had hardly moved from the point whence they had advanced to annihilate the Russian armies and to capture Warsaw." Only the first part of this criticism is correct. While it is true that the Austrians were not yet ready for battle and that the Germans were too weak, the critic overlooks the fact that the decision did not lie with either. Their movements were dictated by the Russians, who had the stronger forces at their disposal. If the German supporting forces had refrained from the advance on Warsaw and had confined themselves to the defensive, the Austrian army would have been exposed to certain defeat.

Ludendorff commanded five Army Corps; the general conduct of the war rested with General von Falkenhayn. It was he who had to determine whether the battle of Ypres was to be continued, and whether it was still worth while to try for a decision in the West. At Falkenhayn's request, Ludendorff attended a conference at Berlin, where it was decided that the centre of gravity was to remain in the West, that the battle of Ypres would be continued, and that Ludendorff must not count upon receiving considerable reinforcements. Yet the position in the East was serious. Once he was assured that the main forces of Germany were engaged in the West, the Grand Duke became audacious. Forty-five Army Corps

THE RISE TO SUPREMACY

were set in motion against Silesia, the point where the Austrian and German flanks touched. Ludendorff was compelled to help himself as best he might. Indeed, he had realised some time ago that it was necessary to make a "new, great resolve". The new plan was to entrain a large part of his army for the regions of Thorn and Hohensalza in the north, whence he intended to advance on the left bank of the Vistula towards Lodz and Lowicz, thus taking the enemy in the flank in his advance towards the west. In one respect Ludendorff's visit to Berlin had not been in vain; he had succeeded in obtaining unity of command in the East. Hindenburg was appointed "Commander-in-Chief in the East" and Ludendorff became his Chief of Staff. The command of the Ninth Army was given to General von Mackensen.

The task with which Ludendorff was faced was one that would have filled any other General with dismay. In spite of every attempt to form new units of garrison troops, untrained soldiers and *Landsturm*-men from the garrison towns of Silesia and Posen, the superiority of the "Russians" could not be made good, and the "Steam Roller" continued steadily to advance. Five and a half Army Corps were gathered for an attack on the Russian flank; but for the defence of the frontier and for the extension of his own front up to the Austrian flank Ludendorff's forces were quite inadequate. The Eighth Army had to retreat before vigorous Russian attacks, and the frontier districts of Eastern Prussia were once more exposed to the enemy, while the whole district east of the Vistula was in grave danger. On the 11th of November, General Mackensen commenced the new operations. The flank attack upon the Russians was a surprise; the flanking movement upon Lodz developed successfully and, with the enemy's rear threatened, a great victory seemed to be in sight. Already General Scheidemann was thinking of evacuating Lodz, and it was only the resolution of the Grand Duke that prevented a general retreat. By his orders the Russian forces were rallied at Warsaw and a new advance was initiated, while fresh divisions arrived to support the Russian resistance. The German forces which had threatened to encircle the Russians were threatened with a similar fate themselves, and one

German Reserve Corps and one Guards Division were in fact completely surrounded. The Russians sent out wireless messages ordering sixty trains for the prisoners they expected to make.

During the night of the 23rd-24th November the German forces broke through the threatening Russian ring and captured 10,000 prisoners. Gradually a definite front took shape, against which the Russians delivered their assaults in vain. The engagements lasted until the middle of December. Although Ludendorff had failed in his aim of inflicting a Cannae on the enemy, the campaign led to the exceedingly valuable result of putting a stop to the Russian advance.

This part of the Eastern campaign, like the first part, has not remained immune from criticism. It is contended that the germs of failure were latent in Ludendorff's attempt to produce a battle of annihilation as preached by Schlieffen. On the other hand, General Hoffmann calls this campaign the best operation of the whole war. The advance on the Vistula in order to give air to the Austrians, the withdrawal to Czenstochau, the change of front on Thorn, and the assault upon the right flank of the pursuing Russians are, in his opinion, strategical operations of a higher quality than any of the victorious battles of the East. That Falkenhayn failed to see the chance of a sweeping victory is, in his opinion, one of the lost opportunities of the war. If he had broken off the battle of Ypres and resolved upon an operation on a really great scale in the East, success would have been certain: two or three fully trained fighting corps would have sufficed to give him Warsaw, and with it the main line of communication of the Russian army. "On the occasion of Colonel Tappen's visit to Headquarters at Posen", he records, "I nearly went on my knees to him, in his railway compartment, asking him to let us have at least two Army Corps besides the reinforcements which he had promised to the forces of the East; Tappen declined. . . . He did not possess the vision to appreciate the consequences of victory."

We have not here discussed the feelings of an ambitious General who sees but is unable to amend the inefficiency of General Headquarters. A victory in the West would have justi-

THE RISE TO SUPREMACY

fied Falkenhayn's strategy. In the East, where it would have been possible to win a crushing victory, the necessary forces were lacking. Ludendorff was not unaware of the impending danger. The Grand Duke, having failed to pierce the centre of the Austro-German forces, would be certain to take the line of least resistance and to try and attack on the flanks. The road to Berlin was barred. It would now be his aim to open a road over the Carpathians to Budapest and Vienna. General von Conrad, fearing an invasion of Hungary, asked for German reinforcements, a request in which Ludendorff supported him and to which Falkenhayn acceded. As a result, the German army of the South was formed, consisting of three German and four Austrian divisions under the command of General von Linsingen. To Ludendorff the post of Chief of Staff of this army was assigned. On the face of it there seemed to be valid reason for this transfer, since it was Ludendorff who had advocated the operations in the direction of Przemysl. However, at Hindenburg's headquarters the impression prevailed that Ludendorff's transfer was principally inspired by a desire to separate him from his chief. This was a second attempt in this direction, and the suspicion might well be felt that the popularity of the two Commanders in the East was not viewed without some displeasure in certain quarters. Hindenburg wrote a letter to the Kaiser requesting him to allow Ludendorff to stay by his side.

Meanwhile Ludendorff had proceeded to Munkacz in Hungary, where the German Southern Army had made its headquarters. These frontier regions, with their mixture of different races, and the whole district in which the new army was advancing, afforded Ludendorff a novel view of the Dual Monarchy. "While I was walking through the wooded hills I approached a sentry who reported to me in an unknown language. The Austrian officer who was accompanying me failed equally to understand him. This gave me an insight into the difficulties with which this army has to contend. . . . The impression I formed was that all except the ruling races were extremely backward. On one of my journeys I passed through the Huzulian villages, and I shall never forget the kind of hovels in which these unfortunate people had to live. . . .

Austria-Hungary is guilty of terrible sins of omission. We, as their allies, should have been able to prevent these."

This simple object-lesson which was afforded to Ludendorff within a few hours' railway journey of the German frontier led him to the following conclusion: "It was a disaster for Germany that we had taken for allies decaying States like Austria-Hungary and Turkey. A Jew at Radom told me that he could not understand how a nation as full of vitality as Germany could tie itself to a corpse." And he concludes by saying: "The Jew was right".

The incidents of the sentry with his unintelligible language and of the Jew at Radom contained more than is implied in Ludendorff's words. Was there anybody in Germany who really knew the countries ruled by the Habsburg dynasty? Had any person in a responsible position looked behind the screen of reports and newspaper articles created at Vienna and hiding the facts from Berlin? Did anybody in Berlin really understand the nature of the allies to whom Germany was tied for good and evil? Mistakes due to ignorance of the most obvious facts cannot be attributed to the malignity of fate; nor is it correct to see in the Austrian alliance the inevitable result of Bismarck's policy. Bismarck himself never treated an alliance as anything except a means to an end, and phrases like that about the Nibelung loyalty would have been banished by him from the regions of practical politics as a sentimental confusion of ideas. Bismarck's idea about an Austro-German alliance implied that Austria was to remain an instrument in Germany's hands, and the Austrian Balkan policy would never have been sanctioned by Bismarck.

The dismay felt by Ludendorff on thus being confronted with the truth about Austria reveals the general ignorance prevalent in Germany. The Prussian General Staff was a perfect military instrument, and there was no detail of technical warfare too difficult for it; but it was unable to extend its horizon beyond calculable magnitudes. Its statistical tables contained numbers of the Austro-Hungarian army and its assumed efficiency; but the military experts lacked the intuitive vision for the nature of things and that happy want of prejudice peculiar to people not overburdened with learning—like the Jew of Radom. The German and the Austrian War Offices had spent enormous sums on espionage in the enemy armies; between friends and allies there was no espionage. The experts in the office of the Engineer-in-Chief knew the enemy armies down to the latest detail of the latest pieces of ordnance; but between the two allies who relied on each other even unto death there was a screen of blind trustfulness. The rigorous exactness of the General Staff, which succeeded in seizing every detail, failed to appreciate the one essential fact that Germany knew nothing about the ally with whom it entered upon the World War. Such is the fate of a purely modern type—the expert. The expert is at once a fanatic in his own subject and completely ignorant of everything that lies outside it: quantity, to put it into Hegelian phraseology, changes its nature and becomes quality. The specialist, a master of his own trade, becomes the slave of his subject, and the security which he feels within the four walls of his department renders him bold also in the face of the unknown. It is this temptation to a misplaced confidence that renders the expert deaf and blind.

The defects of the Habsburg Monarchy were such that they could not be removed by any improvised means. Germany had to pay for its blindness with German blood, and Linsingen's army was only the first among a number of sacrifices. Ludendorff's stay in Hungary soon came to an end. The letter which Hindenburg had addressed to the Kaiser had not been without effect, and by the end of January 1915 Ludendorff was once more at work at German Headquarters at Posen.

The position demanded rapid action. The movements of the enemy, confirmed by intercepted wireless messages, made it clear that the Grand Duke was preparing operations on a large scale. These operations constituted the so-called "gigantic plan", which aimed at a full-dress attack in the Carpathians, coupled with an attack on the feeble northern wing of the German Eighth Army, which was to be surrounded by superior forces and thrust back on the Vistula. Further, strong cavalry forces were to attack between the frontier and the Vistula. Ludendorff's problem was to prevent a wide and comprehensive advance south of the Niemen. He decided to be the first to attack. The General Headquarters had despatched four fresh Army Corps to the East, and it was now possible to aim at a decisive victory. Such a victory, however, could be brought about only by a double outflanking movement. Accordingly, Ludendorff gathered three of the fresh Army Corps on his left wing, whose function it would be to advance from the region of Tilsit and Insterburg, and to surround the enemy, while the remaining and feebler part of his army was to outflank the enemy on the right between Lake Spirding and the frontier. A thin line of troops remained between these two forces, and were to engage the attention of the enemy by a series of attacks.

Every bold operation aiming at a definite decision contains hidden dangers, and the present plan was no exception. The line of the encircling movement was surrounded by a ring of strong fortresses which would remain in the rear of the attacking armies. In the east, on the Niemen, were the fortresses of Kovno, Olita, and Grodno; on the Bobr there was the fortress of Osowiec; in the south, on the Narev, that of Lomza. Post-war critics have blamed Ludendorff for leaving his rear unguarded against possible attacks from these sources. While the four Army Corps of the North were already under way, parts of the Southern (Ninth) Army were set in motion to assist the operation. If the plan was to remain secret the latter movement had to be postponed until the last moment.

The battle commenced on the 7th of February. A snowstorm was raging, and the roads and railways were covered in snowdrifts. The men were faced by superhuman difficulties. Soon after there was a change in the weather and every road became a morass. On the 11th of February the northern wing crossed the road from Stallupönen to Wirballen. On the 14th the southern wing seized the town of Lyck, while the northern wing reached the Augustow forest. On the 18th a considerable part of this surrounding force had reached a position with Grodno in its rear, cutting off the Russian retreat. A ring had been formed, beginning at the German frontier and extending through Lyck in the east, and through Suvalki in the north, whence it passed to

THE RISE TO SUPREMACY 39

Lipsk, twenty miles west of Grodno. The most difficult position was that of the corps commanded by General Fritz von Below. This formation was attacked in front and rear by forces advancing from Grodno and from the Augustow forest. In spite of this, the German ring, measuring no less than 30 miles across, remained unshaken. The Russians surrounded by the ring were compelled to surrender, to the number of 110,000, with many hundreds of guns. Ludendorff fell short of his aim in so far as large numbers of Russians succeeded in escaping to Olita and Grodno before the ring was closed. The weather prevented a Cannae, and Ludendorff was unable to exploit the victory strategically.

After the winter operations in the Masurian districts the Russians made repeated attempts to break the German front before it should be definitely formed. A number of fierce engagements took place along the whole of the Eastern front which made the highest demands on the skill of the generals commanding the weaker forces. The battle did not show signs of slackening until April. Although the Grand Duke's "gigantic plan" had become waste paper, the Russian army was far from being exhausted: the threat on the Carpathian front continued, and it became the task of the German forces to provide relief.

It was at this moment that Falkenhayn resolved to transfer the centre of operations from the West to the East. He was compelled to take this step by the attitude of Italy and the course of events in the Balkans. Ludendorff had advised him long ago to seek a decision in the East, and as far back as autumn, before the winter battle in the Masurian district, he had asked for reinforcements. He tells us that six or eight divisions would have sufficed to carry the campaign to a victorious conclusion "which would have done much to make up for the failure on the Marne". Now that sufficient reinforcements were set free in the West, Ludendorff was tied to the plans of General Headquarters. Falkenhayn with his personal staff moved to Pless in Silesia, and the Kaiser took up his quarters in Prince Pless's castle. Hindenburg and Ludendorff remained at their headquarters at Lötzen.

Ludendorff's whole mentality, coupled with his belief in

the principles laid down by Schlieffen, made it inevitable that he should wish to plan the impending campaign against Russia on different lines from those actually followed by Falkenhayn. While Ludendorff aimed at an outflanking movement on a grand scale, and saw a kind of glorified Sedan before him on the map, Falkenhayn went back to the ideas which he had discussed with Conrad at the Adlon Hotel in Berlin. He was no disciple of Schlieffen's, and he did not hold that one particular form of warfare or one particular manner of seeking a decisive battle could be applied in every case. His calculations were based upon figures; with prudent economy he did not attempt to go beyond what the means at his disposal allowed him to reach, and there was no room in his scheme of things for *imponderabilia* like boldness, tenacity, and luck. The disciples of Schlieffen were continually using the term "annihilation", which for them was the end and meaning of every military action: Falkenhayn avoided this word. For his own method he used the description of "strategy with limited aims", while at the same time he declined to have it described as a strategy of exhaustion—an expression coined by Hans Delbrück in order to distinguish the strategy of Frederick the Great from that of Napoleon. It may be mentioned that Schlieffen's school never recognised this definition; they saw a contradiction in an attempt to look for a decision by avoiding one. Falkenhayn says, indeed, that his method of warfare "never lost sight of its ultimate aim, which was to break down the will of the enemy, any more than it forgot that this aim could obviously be attained only by means of an attack . . . and not by defensive methods"; which caused Ludendorff to remark that it really was difficult to get a clear view of Falkenhayn's ideas about strategy.

At any rate, his aim was not a Cannae; he refused to divide his forces—which a flanking attack would necessitate—and preferred to keep them together for a single assault. Such an assault could only be delivered at the centre. The breakthrough began on the 2nd of May, near Gorlice, where the front made a sharp bend, and proved a complete success. The artillery preparation had been excellent. It had been Ludendorff's task to engage the enemy's attention by a number of

THE RISE TO SUPREMACY

secondary actions, and to draw his attention from the point where the real danger threatened. He did this by means of a gas attack delivered on the front of the Ninth Army, by means of an advance on Suvalki, and by an invasion of Northern Lithuania and Courland. The latter operations have become famous. Ludendorff himself only expected a tangible result from the bold operation in the North. Yet it is precisely the advance into Courland which has been attacked by later critics as a wasteful and meaningless action without serious import. The blame for it has been given to Ludendorff. "At a decisive moment in European history, and at the place where the Russian Revolution had its origin—between the Dukla Pass and Tarnow—no less than ten divisions which we sorely needed were employed upon a military extravaganza." According to General Hoffmann's memoirs, however, it was General Headquarters which was "directly responsible" for the raid into Courland. "In March the Supreme Command enquired whether a cavalry raid on the Kovno railway could be carried out. Two Cavalry Divisions were set aside for this purpose, and these arrived in the middle of April." The undertaking may have done harm by causing extravagant illusions and wishes at home, and, on a later occasion, they gave rise to the formation of new kingdoms and plans of annexation. But Ludendorff himself expected no other than purely practical military results from the raid, and even these only in the event that Falkenhayn should finally decide to attempt an outflanking movement in the North.

The break-through at Gorlice was a great success; but although the Russian front was forced back, there was no overwhelming defeat nor yet any increase in the booty. The operations were approaching their dead centre. Ludendorff's staff, with its eyes ever open for the opportunity of an outflanking movement, seized upon the fact that in the whole of Europe there was only one enemy wing exposed to such an operation. This was the right wing of the Russian forces, and precisely this wing faced the Commanders-in-Chief in the East. "Perhaps for the last time", Hoffmann, a member of Ludendorff's staff, records, "we had the opportunity of inflicting a decisive blow upon the Russian army." The attack would have to be planned so as to cut off the Russian centre, which still formed a salient pointing west beyond Warsaw. At Ludendorff's headquarters there was a difference of opinion. In Hoffmann's view, a comprehensive operation towards the north and east should be undertaken in order to cut off the Russian centre, while Major von Bockelberg was in favour of an offensive beyond the Bobr. Ludendorff supported Hoffmann. His plan was to carry out an operation in an easterly and southerly direction against the fortress of Kovno; having captured this point, on which the Russian lines of the Niemen were based, he would open the road to Vilna and cut off the Russian retreat. Slight reinforcements at the right moment would suffice to deliver a flank attack on the retreating Russians and to inflict a crushing defeat.

Work had already begun upon the preliminary plans when Hindenburg and Ludendorff were commanded to appear before the Kaiser at Posen. The Kaiser, under the influence of Falkenhayn, had rejected Ludendorff's plan. Faithful to his strategy of limited aims, Falkenhayn declined to enter upon an offensive "into the blue". Colonel Bauer mentions that the Kaiser was "wholly under the influence of Falkenhayn and Tappen", and records a brief episode in this connection. In support of his plan, Ludendorff had advanced the argument that an attack in the direction suggested by him would meet with no more than a feeble resistance, whereupon Tappen remarked sarcastically: "These people only want to attack where there is no one to oppose them". Colonel Tappen, the Chief of the Operations Section, is treated as the evil spirit who prevents every plan on a really large scale.

Ludendorff was not a little distressed at this victory of the strategy of limited aims. But the Kaiser had given his command, and in place of Ludendorff's outflanking manœuvre a break-through on the Narev was undertaken by the armies of Gallwitz and Scholtz. The attack commenced on the 13th of July, and was carried up to the Narev; the ultimate aim, however, which was to seize the flanks of the Russian centre by a southerly advance from the Narev and a northerly advance from the Bug, was not attained. Still, the Grand Duke evacuated the entire salient between Narev, Vistula, and Bug,

THE RISE TO SUPREMACY

together with all the fortresses contained in it, in an attempt to save the Russian centre: a striking success had been reached, a holiday was decreed at home, and it appeared for the moment as though Falkenhayn's strategy was vindicated.

Although Ludendorff did not renounce his plan of attacking the Russian retreat via Kovno and Vilna, the middle of August had come before Falkenhayn gave up his resistance to this plan. A military plan, however, is not valid indefinitely; its validity depends upon the time and manner of the execution. Before the break-through at the centre, Ludendorff had recommended an outflanking movement on a large scale, at a time when it still appeared possible to cut off the enemy's line of retreat. It was questionable whether this plan could still be realised. Each day the Russians moved further east, while the forces at Ludendorff's disposal were considerably reduced. Falkenhayn gave permission to draw the investing troops from Novo Georgievsk and to take a number of divisions from the armies of Scholtz and Gallwitz, which by now were concentrated on a narrow front. The bulk of the troops thus set free was sent to the West front and to Serbia under the command of Mackensen. Before this operation Ludendorff had commenced the siege of Kovno with no more than two batteries of his heaviest artillery; on the 18th of August the fortress fell.

Now that the end of August was near, Ludendorff knew very well that if any attack could be made upon the flank of the retreating Russians it could only be done in the general direction of Kovno, Vilna, and Minsk. The attack was entrusted to the Tenth Army under General Eichhorn, while the northern flank of the army of the East (the army of the Niemen under Otto von Below) continued to advance upon Dünaburg. This operation could not be carried out unless the Tenth Army was reinforced. The transfer of the investing troops from Novo Georgievsk and of the division set free at the centre was a lengthy process; the railway could not carry much traffic, the roads were bad, and the horses practically worn-out. Before the reinforcements arrived a heavy attack was delivered on the Tenth Army from Vilna. The Russian centre meanwhile withdrew in good order, and Russian Headquarters even were able to transfer troops from the centre to either wing. Ludendorff noted in his diary: "We are again having a very exciting time".

On the 9th of September the advance could at last begin. General von Eichhorn and his Chief of Staff were full of hope; Hoffmann remarks: "They infect General Ludendorff with their excessive optimism". A break-through was effected, the Russians were compelled to evacuate Vilna, and Ludendorff's cavalry succeeded in reaching Smorgon. At this point, however, the operations, which had begun so auspiciously, met with a resistance which proved insuperable. The Russians once more succeeded in preventing the threatened outflanking movement, and even brought reinforcements from Poland. Nevertheless, the German forces were still in a position to advance west of Smorgon and to reach the district of Baranovitchi and Pinsk and the westerly course of the Beresina. "During the gradual advance from Vilna to Smorgon", Ludendorff tells us, "I realised that the operations would have to be broken off." A further offensive was out of the question: it was time for the army to get ready for winter.

The success of the Russian campaign of the summer of 1915 has obscured the fact that in its essence it is a frustrated plan, and this despite the fact that in the course of six months Russia lost three million soldiers, and guns and stores approximately equivalent to what she possessed at the outbreak of war. In the person of the Grand Duke Nicolai Nicolaevitch, who was relieved of his command, she also lost a commander. Ludendorff and other disciples of Schlieffen consider that a campaign on the lines of their Master would have been more fruitful. When in September the attempt was made to surround the right wing of the Russian forces, the Russian army had retreated so far east that part of these forces could proceed on foot to the point east of Vilna where the German forces were threatening to outflank the Russian army: in July the Russians would have been unable to repulse such an attack even by making use of the railway. A part of the retreating Russians would have had to face the German outflanking forces, while another part would have been thrown

THE RISE TO SUPREMACY

south, where they would have borne the brunt of a frontal attack by the German and Austrian forces which were advancing upon them. Count Szernin has expressed the opinion that it would have been possible to conclude a peace after the battle of Gorlice, and Ludendorff completes this opinion by observing that it might have been possible if Falkenhayn had decided to adopt the idea of a powerful outflanking movement while there was time—that is, before the end of July.

Ludendorff's plans were designed on a grand scale; but, like those of Schlieffen, they required extremely powerful forces on the outflanking wings. Critics after the event have found particular fault precisely with the great dimensions of Ludendorff's plan. They urge that the regions which he tried to cover were far too vast to permit him to win a second Sedan with the means at his disposal. Thus Hans von Hentig insists that the idea of an outflanking movement on an exaggerated scale was due to the self-deception of a man whose one desire was to assume the part of a great soldier, and who never had a proper appreciation for one among the many facts on which he based his calculations, namely, the limitations of fighting and marching soldiers, cold, hungry, and deprived of sleep. He further adds that Falkenhayn's more modest plan, which the Kaiser sanctioned, might have been more successful if the idea of outflanking had been kept alive during this smaller operation. If the break-through of the armies of Gallwitz and Scholtz on the Narev was to be fully exploited, it was incorrect to make a frontal attack on the Russians at the Vistula salient and at the same time to thrust them outside the encircling German forces.

In considering the Russian campaign of the summer of 1915, the strategical purists deplore the fact that the "unity of thought and action" was missing. They desire a triumvirate like that of 1866. Here, however, their premise is wrong: the unity whose absence they deplore was present, and the Kaiser, the Chief of Staff, and the Chancellor were at one. The only parallel between 1866 and 1915 consists in the fact that in King William's headquarters Moltke's plans of encirclement were received with the same coldness as Ludendorff's at Pless in 1915. The course of history is determined by invariable rules based upon human nature, and not by formulæ.

There is the greatest possible difference between the battle of Königgrätz as planned by Moltke and between the actual course of the battle. What Moltke had in mind was completely to surround the Austrians. The Second Army, advancing from Silesia under the Crown Prince, was to take up a position on the left bank of the Elbe, taking the fortress and the enemy army in the rear east of Königgrätz. The First Army was to extend its right wing, consisting of parts of the army of the Elbe, across the river. Benedek met Moltke more than half-way. He took up a position in front of the Elbe and awaited the decisive battle with the river in his rear. If Moltke's suggestions had been followed, the Austrian army would simply have been taken prisoners: the Prussian wing on the left bank of the Elbe would have made impossible a retreat by way of Königgrätz, and not a single Austrian would have crossed the river. There would have been no retreat to Olmütz, and no march on the Danube. Unfortunately, Moltke's plan was not approved by Prince Friedrich Karl, the Crown Prince, or the King. The left Austrian wing was not outflanked, and the Crown Prince effected his change of front on the right bank of the Elbe. Schlieffen's comment on the mistakes of 1866 is that on two occasions the opportunity of a Cannae arose; but the idea of a complete surrounding and annihilation of the enemy was so far from the ruling minds at Prussian headquarters that Moltke's simple and splendid plan was doomed to remain inchoate. The Prussian Commanders contented themselves with forcing the enemy to retreat—exactly as happened in Russia in 1915.

Another similarity between 1866 and 1915 consists in the fact that the battle was successful enough relatively to prevent a serious self-examination. The victory was celebrated: Te Deum was sung: and no thought was given to what might have been. In 1866 the officer commanding the fortress at Königgrätz actually shut his gates in the face of his own troops. The beaten army was held up before the moats and the river; yet the Prussians did not see the need of carrying their victory to its final conclusion. When the pursuit was

THE RISE TO SUPREMACY

eventually taken up, it was carried out according to the instructions of headquarters and not in the spirit of Moltke. It was in vain that the latter pointed out the similarity between the present position and that of Napoleon at Jena, and that, exactly as Napoleon might have done, he gave orders for a direct advance upon the Danube. The army of the Crown Prince was closer to Vienna and Pressburg than the main body of the Austrians, and Moltke's orders could not be executed, because they were not understood. It was this incomprehension which had rendered a battle of annihilation impossible, and which now rendered impossible an equally deadly pursuit.

One of the motives which urged Bismarck to conclude a rapid peace, besides the desire to avoid an enduring hostility with Austria, was the fact that Königgrätz was not truly decisive; the war threatened to drag on, and at any moment France might intervene. One difference between 1866 and 1915, however, must not be overlooked. While Schlieffen is right in saying that Moltke's Generals failed to understand his ideas, Ludendorff could not make a similar claim; for the Army Commanders of the East were fully capable of conducting a war according to Schlieffen's ideals. This much has been demonstrated by the battles of pure annihilation. The obstruction came from Falkenhayn, who played a part in 1915 similar to that played by Prince Friedrich Karl in 1866. On the other hand, William the II. acted exactly as did his grandfather William I. The latter refused to entertain Moltke's idea of a Cannae, the former Ludendorff's idea of a battle in the East inspired by Schlieffen.

Throughout the literature of the war there is to be found the assumption that the triumvirate composed of King, General, and Statesman assured the happy event of Moltke's wars and was a guarantee of victory. This assumption is a pleasing fiction. If all that mattered was harmony between the three decisive men, then the triumvirate of 1915 would have been far superior to that of 1866. Again, no light is thrown upon the problem of leadership if we follow Groener and Schwertfeger in assuming that it is a matter of Destiny whether, in the crucial hour of war, a nation produces an efficient general and an efficient statesman, or is ruined by inefficient leadership. Bismarck and Moltke alike would have refused to have anything to do with an appeal to Destiny. Bismarck was not the man to allow Destiny to play a part in his calculations, and, when decisive steps were being taken, to allow man to appear as the toy of the unknown, the uncertain, and the incalculable; and equally Moltke was far too lucid a thinker to excuse the infirmities of the leader at a crucial moment by a comprehensive appeal to Fate. In one respect both were Prussians in the best sense of the word: they measured the value of the man who feels himself called upon to lead by his capacity, his efficiency, and his success, and by nothing else. There is a certain austere strength in the fact that they lacked any sense for the heroism of the vanquished and for the "tragedy of defeat". It was a later school of historians which made of the Shakespearian drama of Prussian history a pious pageant for family use. In this version no trace is to be found of the real qualities of the master spirits of that time; of their loneliness, their insubordination and rebelliousness, and of their isolation; no trace of their cunning and their astuteness. Their intellectual courage and sense of responsibility have become Sunday School virtues. This picture of them has pleased the multitude. The German subject could not support bare historical austerity; he had to make heroes in his own image before he could understand and applaud them.

In this Sunday School version of history, Destiny takes the place of self-reliance. If Bismarck had been asked what he would have done in the case of a war on two, three, or four fronts, he would have replied that he would never have placed himself in such an awkward position as to be compelled to take refuge in "Destiny". And Moltke would probably have said the same. He differs from modern philosophers of war and philosophising warriors in this, that he never held that mystical view of history where war is interpreted as a vast catastrophe uncontrolled by human will. The two wars which Moltke waged were political acts; the deliberate continuation of politics by warlike means. They were calculable and they had been subjected to reasonable calculation. In 1866 Prussia

THE RISE TO SUPREMACY

could not win the hegemony of Germany except by means of war; but the stakes could be approximately calculated, and, after taking into consideration every possible danger, the leaders were still in a position to balance the stakes against the possible gain. War was deliberately used as a means to enforce the will of Prussia upon an enemy who was not amenable to any other argument. The Franco-German War, although a greater undertaking and of longer duration, did not go beyond the sphere of rational thought. If it had been suggested that a day might come in Germany when a war might break out in some unforeseen shape, like an incalculable event of Destiny, Bismarck and Moltke alike would have relegated the suggestion to the region of dreams and nightmares. There is indeed a famous passage in Moltke where he conjures up the vision of a war lasting five or even thirty years; but this passage has been misinterpreted by the apologists of the idea of doom and destiny in politics. Moltke does not say that this vision of a long and incalculable European War is inevitable; what he does say is that, once this danger has in fact become inevitable, a disaster will have arrived. The essence of Moltke's argument is that when this stage has been reached war ceases to be a political act—the continuation of political intercourse by other than by political means.

The skill of statesmen has always manifested itself in avoiding the dangers of a war on more than one front, either by breaking up a hostile alliance by political measures or by forcing the allies to enter into war singly, or else by assuring themselves of the assistance of allies of their own who undertook the task of attacking the hostile allies and of dividing their forces by involving them in a war on two fronts. In his *Psychologische Strategie des Grossen Kriegs*, Hans von Hentig tells us that a war on two fronts is the unhappy offspring of a mishandled political situation. He treats it as the last and desperate attempt to seek a military remedy against a complete political catastrophe: to be compelled to enter upon a war upon two fronts is itself a defeat. War is the creature of politics; but it is politics which, as Clausewitz says, is "the intelligence"; war is the instrument of intelligence, and intelligence is not the instrument of war. The General Staff would have had a perfect right to question an "intelligence" which instructed it to accept as obvious and inevitable those worst possible conditions of war which had been Moltke's nightmare. Since Bismarck had been deposed, it had been the task of the General Staff to allow for a maximum of dangers; but it would have had to possess superhuman skill to make good all the errors and blunders of German foreign politics by military safeguards. It is a curious fact that the disciples of Clausewitz have been so tolerant as to call this task a "Destiny". Bismarck, with his dislike of abstract thinking, has expressed with his own admirable brevity the essential point in different words. Every political activity, he says, rests upon supposition; we weigh a number of probabilities, and we base our own plans on these calculations about our opponents; if all goes well we win laurels, and if our plans miscarry we get the reputation of incompetence. These remarks were made in 1890 before casual visitors, and there is a certain irony in their assumed modesty. The essence of this simple definition consists, not in the importance attributed to accident, but in the contrast made between intelligence and incompetence. There is no idea of Destiny, and the supremacy of reason remains unimpaired.

It has further been said that Clausewitz is guilty of a confusion of thought, his idea of an "absolute war" being akin to, or identical with, the notion of a war as some cataclysm or stroke of destiny. However, his book *Vom Kriege*, a lucid formulation of political action in general, never mentions an "accidental" war—which would be an abnormal case—because he holds that an irrational war cannot be the subject of rational thought. Clausewitz maintains throughout that war is simply an instrument of politics, and since it is a part of politics, it assumes a political character. "As politics grow in scope and magnitude, war grows with it and can reach a height where it attains its absolute form." This absolute form is reached when the entire force and all the means at the disposal of the nation are placed at the service of war, and the two belligerents are compelled to draw on their last resources. The first instance of an "absolute war" Clausewitz's contemporaries saw under Napoleon. This kind of warfare stops at no sacrifice in order to defeat the enemy, as opposed to the wars of the 18th century, which confined themselves to the art of manœuvring. Clausewitz's definition of "absolute war" has nothing to do with the vague idea of some absolute warlike force beyond the scope of human will, from whose dark centre the thunderbolt is discharged upon the earth.

The form which the war assumed towards the end of 1915 would surely have seemed incredible to Moltke and his staff. The summer campaign in the East led to a political success in the shape of the adhesion of Bulgaria; but it led to no defeat of the enemy. Again, the Serbian campaign was successful, but not decisive. The scene of the war grew more comprehensive and extended south and south-east, embracing Italy, Serbia, Greece, and the whole of Turkey. The burden of the Russian and the Serbian campaign had been borne by Germany, whose troops supported the Austrian front and were engaged in Asia. In the West the resistance grew. Conscription had been introduced in England, and a new Russian army was advancing westwards. In this position Germany's policy was still uncertain, and the war, lacking as it did political guidance, grew more and more "intense" in Clausewitz's sense. Strategically, Falkenhayn supported the ideal of sober calculation and wise restraint; but in the political sphere there was nobody to transform these principles into practice. Clausewitz praises Frederick the Great because "although he pursued an ambitious aim with limited forces, he undertook nothing beyond these forces, and enough in order to reach his goal". But such praiseworthy conduct requires political aims and political thought, and it was precisely in these that Germany was lacking.

Ludendorff remained faithful to his plan of crushing Russia before undertaking an extensive operation in the West. His new plan favoured a movement against the southern Russian wing. This plan also contained a political element; it was hoped to force Roumania to declare herself. Conrad, on the other hand, desired a campaign against Italy. Falkenhayn rejected both plans in favour of his own idea, which consisted in the attack on Verdun. The true motives underlying this plan will always remain unknown: its author is the only person to answer the question, and when he was looking back upon his career, his views were clouded by misfortune. Perhaps Falkenhayn wished to capture the strongest of the French fortresses; perhaps his experiences in Belgium and Russia had inspired him with optimism; or perhaps the attack on Verdun was part of a strategy which undertakes nothing save what the available means allow. Falkenhayn's own later explanations would make it appear that the plan was due to a policy of moderation; the German forces did not suffice for a break-through on a wide front, and accordingly the attack on Verdun was selected as a means of slowly consuming the enemy's forces. But such a strategy would have been reasonable only if backed politically, and if limited strategical aims had been accompanied by a readiness to enter into negotiations. Precisely at this time, however, Bethmann Hollweg said: "After these gigantic events there can be no *status quo*".

Although the Verdun plan is contrary to the principles of Schlieffen, it has not been condemned out of hand by his disciples. Ludendorff himself has expressed the opinion that if the attack of Verdun had been undertaken with stronger forces "it might have had very considerable local and political results". According to him, Falkenhayn's mistake consisted in the fact that he merely aimed at the attrition of the enemy and not at a complete break-through, and in his opinion the obstinate pursuit of the less important aim eventually proved more expensive than a feasible plan with really ambitious aims. Ludendorff's criticism is correct; but it is not merely a matter of moral courage, as Hans von Hentig calls it, if a general recognises the erroneousness of an operation built on false assumptions, and accordingly breaks it off; it is not merely character that is required, but appreciation of facts; and this requires intelligence. Falkenhayn probably did not so much lack the courage to admit openly that he had been wrong; what he lacked was that dialectical manner of thought which Clausewitz had inherited from classical German philosophy, and which would have allowed him to perceive that the costly

THE RISE TO SUPREMACY 53

pursuit of a small aim eventually reaches a point where it changes its nature and grows into a decisive defeat.

Ludendorff was not allowed to take a share in deciding the crucial questions of the war. Falkenhayn remained obstinate in his view that the Russian army had been weakened so much that for some time to come no large scale enemy operations need be anticipated, a view which was inspired by a brief interruption in hostilities, which Ludendorff used for consolidating the German successes. Supplies had to be arranged, a provisional administration had to be set up in the newly acquired provinces, and the resources of the country to be exploited for military purposes. But already in March 1916 it appeared that Russia was not nearly so far exhausted as Falkenhayn had assumed.

On the 16th of March a barrage was put down on the front north of Smorgon of an intensity hitherto unknown on the Eastern front. More than a thousand guns, most of which had been brought to Russia from America via Japan, concentrated their fire upon the narrow strip where General Ewert, the new Russian Commander-in-Chief, had decided to attempt a break-through. The district in question was between Lake Naretch and Lake Wishniew, where it was desired to break the thin line of the German army of the East and to force a road to Kovno. Russian forces also advanced north and south of the two lakes. The offensive had been carefully prepared, and it was almost impossible to beat it back. Ludendorff had no reserves at his disposal, and 66 German had to face about 400 Russian battalions. All that the Commander-in-Chief could do was to withdraw forces from the points where the danger seemed least pressing, and to transfer them where it appeared greatest; and this, Ludendorff himself tells us, is very difficult "if the Commander-in-Chief is compelled all the time to live from hand to mouth". The offensive lasted from the 16th to the 28th of March, after which it came to a standstill. This was Ludendorff's first great defensive battle.

While the battle of attrition continued to rage round Verdun, and General Conrad was initiating the attack against Italy, two events occurred which changed the whole nature of the war. On the 4th of June, General Brusilov broke through the Austrian front at Lutzk, taking prisoner an entire army of 250,000 men with all its guns, and advancing some thirty-seven miles. About the same time the battle of the Somme commenced in the West (24th of June).

The battle of Lutzk was not only the greatest victory won by the Russians during the war, but it also brought a new ally to the Allied Forces in the shape of Roumania. Conrad's offensive against Italy had to be suspended; and the Italians advanced and captured Goricia. At the same time another offensive was launched by the Allies in the Balkans from Salonica. The second of the above-mentioned events, the Anglo-French attack on the Somme, constituted the first great battle based on artillery rather than man-power; and the blow it dealt to the German fortunes revealed at once the errors of Falkenhayn's strategy and the technical inferiority of the German army. "The penalty had now to be paid for all the faults of the German military administration; the lack of co-ordination between the German and Austrian operations all but led to disaster in Galicia, where German troops had to patch up the broken front, while the Austrians made an unnecessary and unsuccessful offensive against Italy. The operations against Verdun were seen at last for the complete mistake they were." The battle of the Somme was the greatest manifestation of technical superiority seen up to this point, and the German Command had nothing to oppose to it save infantry and machine guns. By the summer of 1916 Germany had everywhere been forced into the defensive.

At the beginning of June, Hindenburg and Ludendorff were called to Imperial Headquarters at Pless. To the Kaiser's request for their opinion on the position in the East, they replied by describing things as they were, without any attempt at disguising unpleasant facts, pointing out the lack of unity of command and the inert condition of the Austrian army, which could be stiffened only by the addition of German units. The discussion remained without result, and it was not until after the defeat at Lutzk that the Austrian Supreme Command admitted an extension of the German authority. Once the Austrians had given way, Hindenburg was Commander-in-Chief from the Baltic to Galicia, while General Seeckt was

THE RISE TO SUPREMACY

attached to the armies commanded by Archduke Karl as Chief of Staff. Meanwhile the Russian Commanders transferred the weight of their attacks to Volhynia and Galicia, and Ludendorff, whose headquarters had hitherto been at Kovno, was compelled to move to a point nearer the scene of battle. Accordingly, he went with his staff into the citadel of the fortress of Brest-Litovsk, which had been destroyed by fire and was entirely deserted.

Roumania had declared war on the 27th of August 1916. On the following day Hindenburg and Ludendorff received orders from the Chief of the Imperial Military Cabinet, General von Lyncker, to proceed immediately to Pless. On the same afternoon they left Brest-Litovsk, and did not return to the Eastern front again. On the morning of the 28th, General von Lyncker had called upon Falkenhayn and told him that the Kaiser had sent for Hindenburg and Ludendorff to ask their advice; on the same afternoon Falkenhayn tendered his resignation.

For a second time it had become necessary to find a new leader for the loyal and devoted German armies. After the battle of the Marne, the public had failed to appreciate that the chosen leader was not the right man, and the truth was only learnt gradually. On the other hand, those who could look behind the scenes did not venture to criticise the rights of the Crown or to question the decisions of the Supreme War Lord. The Kaiser had appointed General Falkenhayn without consulting any other person; and it was not until now that public opinion in general began to criticise the conduct of the war. The continued reports of victory had naturally strengthened the self-confidence of the nation; but even the man in the street had enough military knowledge to distinguish between the achievements of the army and the capacity of its leaders. The German forces had advanced far into the enemy countries: Belgium and Northern France were occupied, Courland and Lithuania had been annexed, the German trenches extended from Riga to Galicia, and the German arms protected the south-east frontier of Austria and were masters of the Balkans. All these facts could not dispel the feeling that the war was too much for the Supreme Command.

Full confidence was extended only to the pair of Commanders in the East. In this respect the military instinct of the nation was at one with the opinion of experts and the majority of the Army Commanders, the various Chiefs of Staff, and the front line officers. At the same time there was the possibility that the existence of these criticisms should remain hidden from the Kaiser. At Imperial Headquarters, William II. was an isolated man. His liberty was more restricted than in times of peace, and he could draw material to form an opinion only from his immediate surroundings: from such men as General von Plessen, General von Lyncker, the Chief of the Military Cabinet, Admiral von Müller, and von Valentini, the head of the Civil Cabinet. The Press, being under military censorship, could give no information to the Kaiser. The Chancellor, Bethmann Hollweg, followed general opinion in advising that Hindenburg and Ludendorff should take the place of Falkenhayn; but the advice of a civilian carried no authority and no conviction.

Eventually, however, the voice of criticism succeeded in reaching the Imperial Cabinet. Its source was in the immediate surroundings of Falkenhayn. Colonel Bauer, an expert on artillery, engineering, and chemistry in the Operations Section of the General Staff, records how unbearable the last part of Falkenhayn's regime was for the more intelligent and individualistic among his officers. From week to week, he tells us, things became more desperate. Falkenhayn, formerly unruffled and superior, was visibly losing his calm and security, while Colonel Tappen, the head of the Operations Section, retained his wonted quiet irony and acted the part of complete detachment. "The rest of us witnessed these events with dismay. One day a junior officer was deputed by the others to approach me and to declare that it was my duty as next in seniority after Tappen to interfere. After some hesitation I decided to approach the Minister for War, Wild von Hohenborn. . . . I accordingly called on him, but without result. I next tried my luck with Plessen. At first General Plessen was indignant at this peculiar step, and it took some time for me to explain that it was only as a last resource that we had undertaken this unsoldierly

THE RISE TO SUPREMACY 57

measure. Finally I seemed to have convinced General von Plessen; the latter, however, first sent for Colonel von der Schulenburg, who had been the Chief of Staff before Verdun. With the latter's steps I am not acquainted, but it would appear that Plessen was powerless in the face of the opposition which came from General von Lyncker. These events took place at Charleville at the beginning of August. A few days later we returned to Pless; the position continued to grow more difficult, and we were in despair. On the 27th of August, when I was walking in the castle grounds at Pless with Freiherr von der Bussche, of the Operations Section, we came upon the Kaiser. He was calm and cheerful, and told us that Roumania would certainly not declare war, that the reports were favourable, and that in any case the maize harvest was in progress at the moment. A few moments later we received news in our office that Roumania had already declared war. Next morning I approached General von Plessen once more, represented to him that the only man who could help us was Ludendorff, and begged him to assist us. Soon after I was informed that the Kaiser had been persuaded: Hindenburg and Ludendorff were sent for."

It was in this way that the German army received a leader.

CHAPTER III

DICTATORSHIP

First Visit to Western Front—Roumanian Campaign—Politician *v.* General—Submarine Warfare

BISMARCK has said that the logic of history is even more exact in its audits than the Prussian Audit Board. Hindenburg and Ludendorff were called to the Supreme Command because the appointment was a necessity; it was the adjustment of a past mistake, and even if the obvious solution had been distasteful to the Kaiser, he would not have been in a position to reject it, since he had neither the power to make his own wishes the sole criterion nor the inner strength to trust his own judgment. The reputation of Hindenburg and Ludendorff in the army, their general popularity, and the unlimited confidence which they enjoyed were such as to leave him no other choice. The Military Cabinet had suggested that Hindenburg was to be described as First Chief of Staff of the Armies in the Field, and Ludendorff as "Second Chief". Ludendorff remarks: "To me it seemed that the title of Quartermaster-General would be more appropriate, since I considered that there ought to be only one Chief of Staff. At the same time I insisted on sharing full responsibility for all decisions and steps taken."

The granting of this request had a far-reaching influence on the future decisions of the two Generals. Constitutionally the Kaiser was the Supreme Commander of the army, with the Chief of Staff assisting him in a consultative capacity. The Chief of Staff, in turn, was assisted by the chiefs of the various departments. The title of Quartermaster-General in its traditional meaning was not an appropriate description of Ludendorff's position. The fact, apart from titles, was that the relationship subsisting between Hindenburg and Ludendorff in the East was now transferred to the general conduct of the war, with this sole distinction that Ludendorff's position was considerably strengthened by the granting of joint responsibility, which placed him on the same footing with Hindenburg. Under the regime of the younger Moltke and under Falkenhayn the fiction of the Kaiser's Supreme Command had still been maintained, and even now the external routine was preserved. In place of the Chief of Staff, Hindenburg and Ludendorff continued to report to the Kaiser and to ask for his assent in important decisions; in fact, however, an intermediary stood between the Kaiser and Ludendorff: between the former as Supreme Commander and the latter as Chief of Staff there stood Hindenburg, who actually translated all plans into action.

The co-operation between Hindenburg and Ludendorff depended upon their mutual understanding and confidence. Hindenburg had completely absorbed Ludendorff's mentality and conferred the sanction of his higher rank upon the strategical plans of the Quartermaster-General. Both politically and in a military sense the two formed a single unit. "I myself", Hindenburg records, "have described my relation to General Ludendorff as a happy marriage. In thought and in action we were entirely in harmony, and the words of the one were the exact expression of the thoughts and feelings of the other. Once I had recognised the high merits of General Ludendorff, I considered it my chief duty to leave, or if necessary to create, a free field for the brilliant conceptions, the immense energy, and the untiring will of Ludendorff."

Immediately upon taking over his new position Ludendorff began to introduce changes in the Operations Section of the Supreme Command. Colonel Tappen, who had numerous enemies, was removed, and Major Wetzell was put in his place. The latter was in love with his profession—industrious, pedantic, and a "useful assistant" to Ludendorff. Unfortunately he had some very peculiar tactical and technical ideas, to which he clung with great obstinacy. Colonel Bauer says:

"Although there was continual friction between him and the rest we kept it to ourselves, because we wished to save Ludendorff annoyance. Often enough I had to give way against my own convictions, and have frequently blamed myself for this weakness." The Organisation Section and the Supplies Section were given to Major von Bockelberg. His function was to submit plans for the supply and construction of field guns, machine guns, and *Minenwerfer* to the War Office, and his duties also included the supply of tanks and anti-tank measures. Bauer says that Major Bockelberg was industrious and devoted to Ludendorff, but, unfortunately, "wholly ignorant" of the scientific aspects.

Meanwhile the problems facing the two Commanders had grown enormously. The military position was all but hopeless. Measures had to be taken to keep at bay 750,000 fresh Roumanian troops, while simultaneously the battle of the Somme continued, and each new attack demonstrated the superiority in men and material of the Allied Forces. Austria tended to rely upon German assistance and required support at the front and behind the lines. The same applied to Bulgaria and to Turkey, who stood in need of money, armaments, and means of transport. The position on the Western front was so critical that it was almost impossible to justify the transfer of any divisions to the East. Ludendorff tells us that at that time he did not fully appreciate the danger of this undertaking, and that if he had he would not have had the courage to weaken the Western front.

Ludendorff and Hindenburg arrived for the first time on the Western front on the 6th of September 1916. It did not take Ludendorff long to appreciate what was needed. Although the war had been in progress for more than two years, infantry tactics on either side—and it was the infantry that bore the brunt of the war—had been adapted to a surprisingly small degree to the increased efficiency of artillery. It was as though the military experts had concentrated attention entirely on the destruction of the enemy and very little on the problem of protecting their own men. It was at the headquarters of the Crown Prince that Ludendorff saw the first steel helmet: on the Eastern front steel helmets were unknown. The introduction was due to Colonel Bauer, who had noted in 1900, during the course of experimental firing on corpses, that extremely small pieces of shrapnel, sometimes not weighing more than a few milligrams, sufficed to pierce a leather helmet and the skull. Bauer would probably have failed to overcome the objection of the War Office to such an innovation as the steel helmet without the assistance of Dr. Bier. The latter gave a vivid description of the results of such injuries of the brain, and succeeded in persuading Falkenhayn, who at that time was Minister for War, to adopt the steel helmet.

Military experts tend to be conservative, and even the experience of battle does not easily convert them to innovations. To escape the effects of artillery and machine-gun fire, the German infantry had taken to the trenches; these were continually made more elaborate and secure, but beyond this, no changes were made for two years. Ludendorff immediately perceived the defects of the Western front. Infantry tactics were too rigid, and excessive importance was attached to the possession of territory as such, with the result that high losses were the rule. This tenacity in holding a sector which frequently was of no strategical importance, without regard to the sacrifices involved, was due to primitive notions of courage, and numbers of officers who could not reconcile themselves to such a view were constantly withdrawn, however valuable their services might be otherwise. Ludendorff came from the East with its mobile warfare; he had not come to accept trench warfare as the only possible method, and he attached too much value to the individual soldier to be indifferent to the losses incurred. His first impressions of the Western front led him to introduce a reform in defensive methods; these views were worked out into a system and laid down in a book of instructions for the entire army entitled "Defensive Warfare". Without these essential reforms in infantry tactics it would have been impossible to resist the great offensive in the West and at the same time to keep the proportion between the German and the Allied losses in the ratio of 2 to 3. In place of a number of trenches compressed in a minimum area, and easily visible to the enemy, an extensive system of considerable depth was worked out, admitting of an elastic method of defence. In this system there was more space, the troops were not tied down to a single point, and the elaborate system of trenches made it possible under heavy fire to take cover in several directions. But the German infantry also stood in need of greater offensive power. Ludendorff saw that its main weapon would have to consist in machine guns, and introduced a light model and special sections of machine-gunners. On every field it was urgently necessary to make good the superiority of the enemy. The German superiority in artillery had been lost, and there was a deficiency of guns, ammunition, airmen, and balloons. Without a great effort on the part of industrialists this increase in armaments was impossible; and it was likewise impossible to carry on the war on the new Roumanian as well as on the old fronts.

Winston Churchill, the most fair-minded of the former enemies of Germany, mentions the spontaneous movement of disgust which inspired the Germans in their campaign against Roumania. The Roumanian campaign was a matter of calculation; the splendid opportunity, so long awaited by Roumania, had come, but had also been lost by reason of their custom of bargaining and procrastination in order to make the adventure safe. In June or July, while the South-Eastern front of the Central Powers had not yet recovered from Brusilov's advance, a strong Roumanian attack might have proved successful. But the favourable opportunity had been lost.

Before Ludendorff could turn against Roumania, the Western front had to be secured. The Bulgarian forces had proved a failure on the Macedonian front, and Ludendorff gave instructions for a new army to be formed. The command was given to Otto von Below, who proceeded from Courland to Usküb. The Austrians required support in the Maros sector, and it was only after this had been provided for that Ludendorff could take steps for the Roumanian campaign. Among the many examples of victorious warfare with inferior forces afforded by the German armies of the East, the Roumanian campaign is one of the most remarkable. The first plan of campaign had been arranged by Falkenhayn and Conrad, according to which Mackensen, who had been commanding the German, Austrian, Bulgarian, and Turkish troops on the Danube and the frontier of the Dobruja since August 1916, was to cross the Danube with these forces and advance on Bucharest. The troops commanded by Mackensen consisted of three weak Bulgarian *Landsturm* divisions, one German infantry battalion, one *Landsturm* regiment, and a few squadrons of cavalry, so that the forces at his disposal could not really be described as an army. A German division and a few Turkish divisions, together with some batteries of German heavy artillery, were on the way by rail; but the railways of Northern Bulgaria were anything but good.

Ludendorff cancelled Falkenhayn's plan and gave instructions for Mackensen to invade the Dobruja. The jubilation of the Allies at Roumania's entry into the war had not died down before Mackensen crossed the frontier of the Dobruja (4th of September) and destroyed the fortress of Turtucaia on the Danube (6th of September). The Roumanians, fascinated by the idea of the booty they expected to make, had sent three armies towards the north, two of them across the Transylvanian Alps and one to the Carpathians. The fourth army had not yet deployed when Mackensen struck it. He proceeded to cross the Southern Dobruja by forced marches, and to occupy the fortress of Silistria. Meanwhile the two Roumanian armies crossed the passes and reached Hermannstadt, the capital of Transylvania, Kronstadt, and Fogaras. They thus might hope to make themselves masters of the old German settlements of Transylvania and the rich plains of Hungary.

Apart from a few Austrian *Landsturm* battalions, the Roumanians had met with no opposition, for the simple reason that the Germans had not yet arrived. The main body, the Ninth Army, was under the command of Falkenhayn. "In this important position", Ludendorff says, "he had an opportunity of proving his military capacity as an Army Commander in the service of the Fatherland." Falkenhayn's forces consisted of the Bavarian Alpine Corps, of one division each, drawn from Courland and Alsace, of Schmettow's Cavalry

Corps, and of certain units of Hungarian *honvéd* and Austrian *Landsturm*. The army adjoining these forces in the north and commanded by General Arz was even weaker. On the 17th of September, Falkenhayn took charge, and on the 21st of September the Bavarian Alpine Corps began to move. In the face of an enemy three times as strong as himself, Falkenhayn followed the tactics of Ludendorff at Tannenberg. The more easterly of the two Roumanian armies was kept in check by Schmettow's Cavalry Corps, while the Bavarians executed an outflanking movement and captured the Roter Turm Pass, by which manœuvre they occupied the enemy rear and his most important line of retreat. Meanwhile Falkenhayn delivered a frontal attack at Hermannstadt against the western Roumanian army. Compelled to retreat until they had their backs to the mountains, this army came within the reach of the Bavarians, whose mountain artillery and machine guns did deadly work from their emplacements above. The annihilation of the first Roumanian army was complete, and Falkenhayn could now turn to the second army. The battle culminated in the struggle for the Geisterwald near Kronstadt. On the 8th of October the Roumanians were driven from Kronstadt, and on the 9th were forced across the Predeal Pass and down the slopes of the hills. The most northerly of the Roumanian armies gave up all idea of aggression after these defeats of the sister armies and sought safety on the Sereth, with the Austrians in pursuit.

Ludendorff perceived that it would be a mistake not to follow up the successes won in these few days. Meanwhile, however, circumstances had changed. The Allied Commanders, alarmed by the German successes, saw the natural resources of Roumania, its oilfields and grain, in the hands of the Germans. In consequence Russia sent troops, France officers, and England engineers to destroy the oil plant. The German reinforcements, consisting of four infantry divisions and two cavalry divisions, did not advance, as the Roumanians expected, by the shortest route across the Predeal Pass, but across the Vulcan Pass further to the south-west. A final battle was engaged in the Walachian Plain, on the left wing of the Roumanian northern front. On the 28th of November vast conflagrations marked the path of the retreating Roumanians. The three German forces began to converge upon Bucharest. This moment had been awaited by Mackensen for days. On the 24th he crossed the Danube, and a day later his cavalry had got into touch with the advance units of Falkenhayn's army. A desperate resistance under General Berthelon at the gates of Bucharest could not save Roumania. On the 6th of December Mackensen's troops entered Bucharest. Ludendorff records how great his relief was when he received the report that the German Cavalry Divisions had found the northern defences of the fortress blown up and unoccupied. The Roumanian campaign had been directed by Ludendorff himself from Pless. There were many anxious days, particularly during the second part of the campaign; but no detail escaped the eye of the Commander. Falkenhayn complained that streams of telegrams, "equally superflous and annoying", flowed from Pless.

Ludendorff was justified by success. "If success had been wanting," he remarked, with reference to the above-mentioned complaints, "opinions about the Roumanian campaign would have been very different."

The sure guidance of Ludendorff had succeeded in overcoming a dangerous crisis. Roumania was defeated, while the Allied losses and climatic conditions made it necessary to break off the battle of the Somme, which had been raging with the greatest intensity up to October. Before Verdun, too, there was a cessation of hostilities. In London and Paris such a conclusion had not been expected. Thus Churchill wrote: "The general misfortune in which the year 1916 closed produced feelings of disappointment and vexation in the Cabinet which overcame this personal misunderstanding. The failure to break the German line in the Somme battle in spite of the enormous losses incurred, the marvellous recovery of the German powers in the East, the ruin of Roumania, brought, as it seemed, so incontinently into the war, and the first beginnings of a renewed submarine warfare, strengthened and stimulated all those forces which insisted upon the need of still greater vigour in the conduct of affairs."

Ludendorff was under no misapprehension. He was aware that the Western forces were completely exhausted, and he foresaw that the enemy would leave nothing undone to increase their men and material. The English army grew in numbers, France drew upon her colonial forces to the fullest extent, and Russia organised a fresh defence, while an army was formed under French officers out of Roumanians and Austrian prisoners of war. Three-quarters of the world were engaged on the equipment of these forces. "It was to be expected", Ludendorff says, "that the numerical and the material superiority of the enemy would be accentuated in the course of 1917.... Our position was extremely difficult, and it seemed impossible to find a way out. We ourselves were not in a position to attack, and we dared not hope that any one of our enemies would collapse. If the war continued for any length of time, defeat seemed inevitable." In spite of these facts, feelings of disquiet and uncertainty were felt in the enemy camp towards the end of 1916. The terrible losses of the French at Verdun and on the Somme—365,000 dead and 579,000 wounded, without counting Belgians and English—had struck a severe blow at the whole nation. In England there was a serious crisis, and Churchill refers to the regrettable panic at home which accompanied the slaughter in the trenches.

The home of parliamentary institutions, if looked at carefully, is found to be going through the same crisis as Germany with its lack of political training and experience. The populace began to mistrust their own Governments and to lose confidence in their political leaders; in England no less intelligent a person than Winston Churchill has placed on record his dismay at this process: "A series of absurd conventions became established, perhaps inevitably, in the public mind. The first and most monstrous of them was that the Generals and Admirals were more competent to deal with the broad issues of the war than abler men in other spheres of life. The General no doubt was an expert on how to move his troops, and the Admiral upon how to fight his ships, though even in this restricted field the limitations of their scientific knowledge when confronted with unforeseen conditions and undreamed of scales became immediately apparent. But outside this technical aspect they were helpless and misleading arbiters

DICTATORSHIP

in problems in whose solution the aid of the statesman, the financier, the manufacturer, the inventor, the psychologist, was equally required. The foolish doctrine was presented to the public through innumerable agencies that Generals and Admirals must be right on war matters, and civilians of all kinds must be wrong. These erroneous conceptions were inculcated billion-fold by the newspapers in the crudest forms."

Asquith was swept aside by this wave of popular panic, and Lord Northcliffe ruled in his stead.

This was not the time to make overtures of peace. To take such a step would have been the work of statesmen, financiers, scientists, and psychologists, and these had been pushed into the background and were afraid to speak. Lloyd George and Northcliffe were the spokesmen of that period.

The Chancellor, Bethmann Hollweg, had been considering in October 1916 the idea of suggesting to President Wilson the desirability of acting as mediator for peace. The Kaiser agreed to this plan. Count Bernstorff, then the German Ambassador at Washington, was to have approached the President. When Ludendorff was consulted in this matter he had made no objections. "I fully concurred," he says, "and I was pleased at bottom, although the known determination of our enemies to crush Germany made me sceptical. Their prospects for 1917 were so much more favourable than ours that I doubted if any steps towards peace would succeed. At the same time I was not quite without hope. . . ." The time had not been chosen well for approaching Wilson, since the presidential elections were pending in Washington. Meanwhile, Baron Burian, the Austrian Minister for Foreign Affairs, suggested a direct offer of peace to the Allies. Here again Ludendorff records: "I lent my support in this step in so far as the Chancellor consulted me". On the 12th of December 1916 a declaration was handed to the representatives of the neutral States in which Germany declared herself ready to enter into peace negotiations; the declaration did not mention any conditions on the part of Germany. Shortly before, Lloyd George had declared that the evacuation of Belgium and Northern France must be the preliminaries of peace. Before the Allies had time to reply to the German declaration, Wilson, who had been re-elected President of the United States, addressed a note to all the belligerents and to the neutral States in which he made an offer of mediation (18th December). He had in mind a new Congress of Vienna, a kind of general meeting of all the Powers, in which, like Metternich a hundred years ago, he would give peace to an exhausted Europe.

What Wilson desired was to be informed by all the belligerent States what were their precise aims, so that he might form an idea how near the world was to the peace which all mankind was so earnestly desiring. In their note of the 26th December, Germany and the other Central Powers evaded this question; they preferred to negotiate direct with the enemy. On the 30th of December the Allies gave their reply to the German declaration, declining to examine such an insincere and insignificant suggestion. On the 10th of January 1917 the Allies informed Wilson that it was impossible at this stage to conclude a peace ensuring the reparations, restorations, and securities which the Allies had a right to claim. Their aims were described as being the restoration of Belgium, Serbia, and Montenegro, the surrender of those districts which had formerly been seized against the will of their inhabitants, and the liberation of the Italians, Czechs, Serbs, Slovenes, Roumanians, and Slovaks from an alien domination. The question was whether these were the real conditions or whether the Allies were exaggerating for fear of appearing weak.

Ludendorff himself was guided by a similar fear. "We had to avoid anything that might have been interpreted as a sign of weakness. Such action would have had a demoralising effect on the army and the nation and would have caused the Allies to redouble their efforts." At the beginning of 1917 Ludendorff had not yet become the sole controlling force in Germany, and his mentality, with its militaristic bias and its dislike of psychological considerations, had not yet begun to exercise a guiding influence upon politics. He was a soldier, and it was not to be expected that he should be skilled in interpreting motives and in reading the political situation. Ignorant of the technique of diplomacy, he had never attempted to appreciate an opposite point of view, and could not know how much fear and mistrust, how much hope and expectation, was hidden behind the brave language of his opponents. At the same time the majority of German politicians, of the Press, and of the whole nation was ready to accept the arbitrament of the new Army Commanders even in matters of policy.

Field-Marshal von Hindenburg has admitted with generous candour his lack of political sense. He says: "My inclination did not lead me to follow the life of politics. I cannot tell whether my power of political criticism was too weak or my soldierly feelings too strong; indeed, my dislike for diplomacy in general is probably due to this feeling. I admit that such a dislike might be called blindness or prejudice; in any case, I would never have denied the fact, even if I had not been compelled to voice it so frequently and loudly. I always had the feeling that diplomacy made demands upon us Germans which it did not lie in our nature to fulfil; and perhaps this is one of the chief reasons of our backwardness in foreign affairs." Ludendorff had no more political sense than Hindenburg, but he differed from the Field-Marshal in seeing a positive advantage, and not a defect, in the limitations imposed by his military mentality.

It is this misconception which is the ultimate cause of all the faults of German diplomacy and the ultimate tragedy of Germany. A politician by his very nature, mentality, and conduct is an object of suspicion to the soldier. Political talent comes from a different source from that of the military expert; it requires a different kind of knowledge of human nature, and a greater knowledge of international factors than a soldier can possibly possess. But it is not true that this talent is necessarily alien to the German spirit; a glance at Frederick the Great, Metternich, and Bismarck suffices to correct such an error. It is true, on the other hand, that the political gifts of Germans cannot be recognised so long as German historians provide us with distorted images of our national heroes. When generals take to politics they are fond of appealing to the example of Frederick the Great and Bismarck, but these men were totally devoid of any primitive and trustful reliance on the justice of their cause or the justice of the sword. Their political superiority was based upon the scepticism with which they regarded everything which is accepted without question by the governed classes. It is due precisely to this profound scepticism that they feared "the gods", in a sense different from the trivial meaning which Bismarck gave to his famous saying when addressing the subjects of his rule. They did not trust "the gods", they feared men, and they relied only upon their own intelligence. It may be said that these are qualities of the Ancients or of the Latin races; in any case, the few men of political genius whom Germany has produced were remarkable for these qualities. The political backwardness of Germany is not a natural deficiency; it was a characteristic of one, but not of every period in its history. England possesses a number of ancient families in which a concern with public affairs is considered a duty which has passed from generation to generation. In France the publicity attaching to political life and the intense political ambitions of the nation have given rise to a large and apparently inexhaustible reserve of political talents. It is incorrect to say that Bismarck's Constitution and the weakness of Parliament have prevented the development of political talent in Germany, for unless a man of genius is born to be a statesman, political talent requires time to mature; and the German bourgeoisie had not been in existence long enough to produce such talent. At the climax of its historical development it had to give way to younger classes, which pushed it aside. The members of the lower middle classes, in turn, required decades in order to overcome the defects and limitations of their birth and to outgrow their own narrow horizon. Stresemann is the classical example. In the decisive years during which the tragedy of Germany was preparing, his political mentality, in spite of all his natural gifts, remained that of a provincial newspaper reader, and he needed a knowledge of the greater world and an immediate perception of the hidden factors in international affairs before he could cast off his past and grow into his own dimension as a figure of European magnitude. A German who is born a European is spared this painful growth.

DICTATORSHIP

In considering the part which Ludendorff is about to play, we might be led to believe that it was his victorious sword which allowed the General to thrust aside opposition until he became the dictator of Germany. Yet two events which he promoted and influenced contributed to bring about the destruction of the German people: the Polish decisions, and the resolve to carry through unrestricted submarine warfare.

The Eastern policy of Germany, this pitiable child of unsuspecting parents, was pursuing in the autumn of 1916 a road which had been determined long ago. Bethmann Hollweg, the man who bears the chief responsibility for this melancholy chapter of German history, attempts to make it appear in his *Betrachtungen* that Germany was acting under a kind of higher compulsion. "Theoretically", he says, "it might have been right to leave the problem in suspense during the duration of the war"; adding, however, that events made it necessary to proclaim the kingdom of Poland on the 5th of November 1916: the events themselves brought about an accomplished fact.

Yet there can be few events in history which more completely fail to embody the mechanistic concept of the pressure of facts. To see in events the compulsion of superhuman forces is a weakness of those who fail to make themselves the masters of historical events; the German policy in the East had no sense of realities and no plans, and consequently it was surprised by the facts. A statesman with a concrete sense of imagination having to conduct a war on two fronts would have allowed no day to pass without looking for the chance of a separate peace in the East. All other considerations would have been secondary to this, and the question of Poland in particular would have played a subordinate part. Not one of the arguments produced by Bethmann Hollweg suffices to convince historical investigation that in November 1916 there was any compulsion on Germany to proclaim the kingdom of Poland and therewith to lock the door to Russia, then already in a state of collapse. "Everywhere in the country", Bethmann Hollweg records, "we met with the idea of an independent Poland, and it was impossible to blind oneself to this idea in the long run . . ."; assigning a secondary importance to the true cause of the fatal German blunder by suggesting that greater weight was given to a sentimental regard for Polish aspirations than to the wishes and threats of the Habsburgs. If it is desired at all costs to interpret mistakes as due to destiny or a higher compulsion, then we may say that this was yet another instance where Habsburg was the destiny of Germany; and certainly Bethmann Hollweg was not the man to sever the wires pulled at the Ballplatz. He was just strong enough to avert the so-called Austro-Polish solution with its incorporation of the whole of pre-partition Poland in the monarchy of the Danube, but not to grasp that an independent Poland in itself was an error, and that the plan to make one must be repudiated.

Although the negotiations between Berlin and Vienna go back to 1915, a formal agreement was not reached until the 13th of August of the following year. "The political situation in Russia was so obscure", Bethmann Hollweg tells us, "that it was essential for the time being to await the course of events. . . . The question was finally raised again by the soldiers." It is no excuse for a political leader to allow the military to exercise a decisive influence. In the summer of 1916 the Chancellor was free to enter on any course that he liked; the Kaiser was too diffident and reserved to interfere with him, while an important section of the Reichstag, consisting of Conservative and National Liberal members, among them Freiherr von Richthofen, the only member with any sense of politics, had political instinct enough to oppose the plan of proclaiming a kingdom of Poland, so that Bethmann Hollweg would not have been alone in following the statesmanlike course. The fact is that he was not sorry to witness the activities of the Generals, because their plan was his plan. The leading part in this agitation was not played by Ludendorff but by General von Beseler, the Governor-General of Poland.

What was the motive that urged the Generals to press on the Polish question and to insist on the necessity of severing Poland from Russia? It was a naïve belief in the possibility of acquiring a Polish army, and it must be admitted that Ludendorff vigorously supported General von Beseler in his advocacy of the plan. He deals with this question quite candidly: "To me it seemed possible that Poland would be ready to sacrifice her sons in order to free herself from Russia. . . . The Governor-General of Poland believed that the plan of forming a Polish army was not only possible but extremely hopeful. The difficulties of the military position were such that the Central Powers were in urgent need of increasing their forces. Consequently it was the duty of General Headquarters to examine the question of a Polish army." The continued failure of the Austrian army made it all the more desirable that fresh forces should be available. In a letter to Herr Zimmermann, then Secretary of State in the Foreign Office, Ludendorff expresses his apprehension at the continued decline of the Austrian army and points out the need of drawing a substitute from Poland. "Let us found a Grand Duchy of Poland . . . and a Polish army under German leadership. Such an army is bound to come some day, and at the present moment we can make use of it." It was a characteristic error of the German military expert to apply his own manners of thought to a foreign country and people, and to work out the number of new Divisions as a simple percentage of the population. Freiherr von Conrad had pointed out long ago that it was in vain to hope for competent troops, loyal to Germany as well as to Austria, from Poland. And indeed this piece of knowledge was obvious enough. A superficial knowledge of the Poles, of their mentality and of their political methods, would have prevented the German Generals from pursuing this hopeless aim. But this blunder of the military experts, of which Ludendorff himself was not guiltless, was not the deciding factor in the Polish policy of the Central Powers.

When Ludendorff came into power the agreement reached between Bethmann Hollweg and Baron Burian was still in force. The plan to proclaim a kingdom of Poland had been shelved in July 1916, but had not been forgotten. In those days Hugo Stinnes had an interview with Monsieur Protopopov, the Vice-President of the Duma, at Stockholm. Even to-day, when most of the relevant documents are available, it is impossible to judge whether the steps taken towards peace in the summer of 1916 had any chance of success if pursued vigorously and with diplomatic skill. In any case, Bethmann Hollweg sought for such a peace, although, as he says himself, "he had not yet obtained full information about the development of Russian policy". At the same time "his past experience ... did not lead him to believe that a separate peace with Russia lay within the bounds of possibility". And elsewhere he tells us: "I never believed that the Stockholm conversations would have a decisive influence on the course of things, and in the event I was proved to be right".

Thus the fact is that Bethmann Hollweg did not seriously believe in the possibility of a separate peace with Russia. This scepticism could be read plainly between the lines of the letters addressed to General Headquarters. Hindenburg and Ludendorff were interested only in the Chancellor's negative attitude, and did not appreciate the doubts which assailed him; but they had a right to assume that the Chancellor shared their opinion. All that they were interested in was to obtain the Polish Divisions, and, unhampered as they were by any thoughts of a separate peace with Russia, they urged the Chancellor to effect the proclamation of the kingdom of Poland which had been prepared long ago. The decision was finally taken at Imperial Headquarters at Pless on the 18th of October 1916. "At this Council", Bethmann Hollweg writes, "I gave my unequivocal assent to the urgent pleas of General Headquarters, the more so since Stürmer's Ministry had not made even the coyest gesture in response to the vague suggestions of peace. So long as the Polish manifesto did not completely remove every chance of peace with Russia, such a measure was not only admissible but essential, since according to the conviction of the military leaders it would increase the number of troops at our disposal. Eventually the political and the military leaders were fully agreed on this question." Here the situation is perfectly clear. Bethmann Hollweg was convinced that now and in the future it was hopeless to look for a separate peace with Russia, and consequently he adopted the arguments of the Generals and agreed with them in voting for the proclamation of the kingdom of Poland. This fact is not altered by the extreme vigour with which the military leaders insisted on their point of view. It is the function of the political leaders to disarm military impetuosity by reasonable arguments and superior knowledge. Bismarck would not have allowed himself to be bluffed by any rattling of the sabre. In September 1915, Bethmann Hollweg had written to Falkenhayn and had brought forward perfectly reasonable doubts against the idea of the military exploitation of Poland. At Pless, on the other hand, not even the attempt was made to discuss the question whether the Polish Divisions existing on paper would ever become a military reality. If it is urged, after the event, that the fateful decision was due to Ludendorff's insistence and obstinacy, it can fairly be objected that in the end the political leaders acted no more prudently than himself. It was the function of the politicians to demonstrate the worthlessness of the military thesis. An old Jew at Radom had already pointed out to Ludendorff the fundamental mistake in the German policy. If the Foreign Office at Berlin numbered among its staff no official with an expert knowledge of Poland, there were dozens of people in Poland and in the immediate surroundings of Pless who might have furnished a convincing demonstration that the belief in the Polish forces was an illusion and could never be anything else.

The kingdom of Poland was proclaimed on the 5th of November 1916. It was not long before the proclamation was revealed as a gross blunder. The Poles accepted the gift as something that was no more than their due, and did not dream of identifying their destiny with that of the Central Powers, still less of furnishing them with soldiers to fight the Allies. Historical justice compels us to add that the Polish leaders had never attempted to disguise the facts. Professor Bredt, a member of the Reichstag and a reserve officer, who had been wounded and in the autumn of 1916 appointed Town Major at Czenstochau, declared in an expert opinion furnished to the Parliamentary Committee that the Poles should not be blamed for the complete fiasco which attended the proclamation, and that "there is no ground for accusing them of treachery towards Germany".

The first visible result of the Polish policy of Germany and Austria was the fall of the Russian Prime Minister Stürmer, a man who, in spite of Bethmann Hollweg's doubts, was more likely than anybody else to entertain the idea of a separate peace with Germany. After the proclamation the Nationalists under Maklakov came into power.

It is easy to understand the bitterness with which Ludendorff regards this fatal blunder. "The Chancellor and General von Beseler", he writes in his *Memoirs*, "supported the formation of a Polish army and the restoration of the kingdom of Poland. There were many authorities in Germany who raised grave objections to this plan, whereupon rumours immediately spread from Berlin that I was its author. I repeatedly requested the Government to make a statement upon the events, but no politician could be found to accede to this request. . . . All those who have collaborated with me are aware that I was always ready to listen to plain speaking and to weigh objections, but that at the same time I insisted on complete candour." Further he says, with pointed reference to the political leaders: "I would have much preferred a peace with Russia to the entire Polish army, the more so as in my inmost heart the fact of my having been born in Posen made the idea of such a kingdom abhorrent to me. . . . The only difficulty was that hopes and wishes in this case did not suffice to bring about a peace."

Was it Ludendorff's opinion or his will that finally decided the issue? The question of the proclamation of the kingdom of Poland was political, and depended upon the other question, whether it was desired or held possible to bring about a peace with Russia. General Headquarters were blind to everything except the Divisions which they hoped to obtain. These expectations were erroneous, but no objection and no proof was ever brought forward to refute them. On the contrary, at the critical meeting the Chancellor concurred with the Generals—the political and military leaders were at one. But this fact is forgotten, and it is Ludendorff who is blamed for the faults of the German politicians. Thus, Dr. Arthur Rosenberg, usually a very fair critic of Ludendorff, declares, in his *Entstehung der deutschen Republik*, that the main responsibility for the restoration of Poland rests with Ludendorff, since his authority would have been enough to put an end to

DICTATORSHIP

the scheme if he had wished. Rosenberg's assumption is incorrect, since it attributes to Ludendorff dictatorial powers which he did not possess in the autumn of 1916. Further, his description of the position at the decisive moment is incorrect. Ludendorff was blinded by a delusion natural in a soldier: in the pursuit of the will-o'-the-wisp of the Polish Divisions he supported the restoration of the kingdom of Poland, which was fundamentally distasteful to him. It was the duty of the political leaders to demonstrate the errors of the military plan, instead of which the Chancellor followed the Generals in their error and supported them in the idea of restoring Poland—a plan whose ultimate realisation had, in any case, been decided upon long ago. Thus all the deciding factors—the Kaiser, the Chancellor, and the Generals—were of one opinion; so that Ludendorff could hardly be expected to resist alone.

Dr. Rosenberg could not but be aware that there was another power which failed to contribute its advice at this decisive hour. The Reichstag is represented by Dr. Rosenberg as being powerless, and he treats the futility of German politics as the product of the confusion and backwardness of German political evolution since 1871. According to him, the whole lamentable period of the peace negotiations and the part played in them by Ludendorff can be explained only as the aftermath of the Constitution created by Bismarck and of the incapacity of the Kaiser—the German people had to suffer the consequences of having received its unity as a gift from the King of Prussia instead of having to fight for it. We shall see that this judgment is incorrect. Whatever may be our estimate of the part played by Bismarck in forming the character and the political sense of Germany, it is certain that the Constitution of the German Empire was not the cause of the political incapacity of the nation. The German Reichstag was far from being the Cinderella of a police state as which it is so often described; nor did the Kaiser possess the omnipotence whose trappings he assumed before the eyes of the world. The sources of what is called the political impotence of the German nation are not to be looked for in the Constitution of the Empire; its sources are of a far more profound nature.

As far as the Constitution was concerned, nothing prevented the Reichstag from playing a lively and active rôle from the first day of war; from attaching importance to, and demanding, full information; from forming an opinion, from asserting the right of criticising, and from exercising this right. It is a widespread but erroneous belief that with the outbreak of war, emergency legislation and regulations were applied to the Reichstag. It is true that the Kaiser had the right of assembling, opening, proroguing, and dissolving the Federal Council as well as the Reichstag, but without its own consent the Reichstag could not remain closed for more than thirty days, and it always lay within its own powers to make use of the right of free speech. Nor was there any intention of infringing upon the rights of the Reichstag. Between the 4th of August 1914, when, according to the Kaiser's own words, there were no parties but only Germans in the country, and 1916, the Reichstag was convoked eight times, after which time no prorogation took place. Doctor Bredt's scholarly investigation of the part played by the Reichstag during the war makes it abundantly clear that nothing prevented the Reichstag from asserting itself. "If the Reichstag at the beginning had possessed a definite majority with a definite policy it could easily have obtained control of the conduct of affairs. That it failed to do so was disastrous for the German people. It would be unfair to urge that the Reichstag was not acquainted with the true position; the Polish question was perfectly comprehensible and there was nothing here to conceal. Nor need we assume that the leading politicians seriously believed in the idea of a Polish army, and if they had so believed a simple enquiry would soon have convinced them of their error. In fact, nothing was done. The decisive political steps were taken without the Reichstag being consulted. Speeches were delivered, and that was all."

The Polish decision was the first of the two most serious decisions of the period. The decision to embark on unrestricted submarine warfare was the second. How amazing was the act which brought into arms against us a more powerful force even than Russia we know from the writings of our enemies.

DICTATORSHIP

Thus, Churchill writes: "If the Germans had waited to declare unrestricted U-boat war until the summer, there would have been no unlimited U-boat war and consequently no intervention of the United States. If the Allies had been left to face the collapse of Russia without being sustained by the intervention of the United States, it seems certain that France could not have survived the year, and the war would have ended in a peace by negotiation, or, in other words, a German victory. Had Russia lasted two months less, had Germany refrained for two months more, the whole course of events would have been revolutionised. Either Russian endurance or German impatience was required to secure the entry of the United States."

Nowadays an unprejudiced Englishman will admit that the full exploitation of submarines cannot be interpreted as an infringement of the law of nations and of the rules of seafaring peoples. Let us quote Churchill once more: "The Germans were newcomers on salt water. They cared little for all these ancient traditions of seafaring folk. Death for them was the same in whatever form it came to men. It ended in a more or less painful manner their mortal span. Why was it more horrible to be choked with salt water than with poison gas, or to starve in an open boat than to rot wounded but alive in No Man's Land? The British blockade treated the whole of Germany as if it were a beleaguered fortress, and avowedly sought to starve the whole population—men, women, and children, old and young, wounded and sound—into submission. Suppose the issues had arisen on land instead of at sea; suppose large numbers of Americans and neutrals had carried food or shell into the zone of the armies under the fire of the German artillery; suppose their convoys were known to be traversing certain roads towards the front: who would have hesitated a moment to overwhelm them with drum-fire and blast them from the face of the earth? Who ever hesitated to fire on towns and villages because helpless and inoffensive non-combatants were gathered there? If they came within reach of the guns they had to take their chance, and why should not this apply to the torpedoes too? Why should it be legitimate to slay a neutral or a non-combatant on land by cannon if he got in the way, and a hideous atrocity to slay the same neutral or non-combatant by torpedo on the seas? Where was the sense in drawing distinctions between the two processes?"

This is the opinion of Winston Churchill. There could then be no two opinions as to the justification of using submarines. It was equally certain that they constituted an extremely effective weapon, but this weapon only became fully effective if it was possible to torpedo an enemy craft without warning. If the commander of a submarine was tied down by orders to warn an enemy mercantile vessel before attacking, the submarine would always be in danger of being rammed or, if the vessel possessed guns, of being shelled. Even the most cautious submarine campaign, directed only against enemy vessels, could not entirely obviate the risk of conflicts with neutral Powers, and especially with the United States. Subjects of such States might be, and in fact frequently were, on board of enemy vessels. Thus the danger of complication with the United States was almost unavoidable, as was proved by the sinking of the *Arabic*, the *Sussex*, and the *Lusitania*. It became necessary to decide between the value of submarine warfare and the dangers which it implied.

From the outset the Chancellor, Bethmann Hollweg, held the view that political considerations must be paramount even in submarine warfare, and that a conflict with the United States must be avoided at all costs. So long as his authority was unchallenged this line of conduct was followed. Admiral von Tirpitz, in his capacity as Minister of Marine, was subordinate to the Chancellor, and when a divergence of opinion arose between the two in 1915 the Kaiser decided in favour of the latter. Thereupon Tirpitz tendered his resignation, which, however, was refused, with a reprimand. Thus, in 1915, the last word rested with the politicians, and the case was not altered in the following year when, in March, General von Falkenhayn was converted to the belief in unrestricted submarine warfare. Here again the Kaiser supported the Chancellor. Admiral von Tirpitz, who had not been invited to attend the Council of the 6th of March, felt slighted once again, and once again tendered his resignation, which on this

DICTATORSHIP

occasion was accepted. Admiral von Capelle, his successor, had to undertake to respect the opinion of the Chancellor in all political questions.

In dealing with these conflicts Bethmann Hollweg was guided by two considerations. On the one hand he was undoubtedly right in pondering the possible reactions in the United States; but he attached even more weight to the conviction which he had gradually formed that in its calculations about and its appreciation of submarine warfare the Admiralty had been guilty of grave mistakes. The submarine campaign began in February 1915; at first 21 U-boats were in use, to which 14 more were added in March. Yet Admiral von Tirpitz and the Chief of Admiralty Staff, Bachmann, gave it as their expert opinion that within six weeks of submarine warfare England would undoubtedly begin to show signs of weakening. Let us now hear the enemy's opinion. We may fairly quote Churchill's views, since Churchill never underrated the efficacy of submarines.

"If Tirpitz . . . had made no submarine attack on commerce until at least two hundred U-boats were ready, and had not provoked us to counter-preparations in the meanwhile, no one can say what the result would have been." "The attack of twenty-five U-boats in February 1915 was absurd."

Tirpitz was of a different opinion; not only did he believe that 25 U-boats would suffice within six weeks to bring England to her knees, but he also voiced this view to the world. An interview which he gave in December 1914 to the Berlin correspondent of an important New York newspaper made clear his opinion that, in the submarine, Germany possessed an infallible weapon which would break the blockade and soon give her a victorious peace. These and similar utterances brought about two results: the enemy was impelled to devise anti-submarine measures, while Germans began to be intoxicated by that belief in submarines which blinded them to the last. Bethmann Hollweg's comments on his efforts to provide an antidote for this intoxication make melancholy reading. "I never ceased to attempt to limit the unrestricted Press propaganda. But what could the Censorship do against the force of a conviction which might appeal to the infallible judgment of the Admiralty? Any attempt to restrict this agitation only made matters worse. . . . I could not state in the hearing of the enemy that the only reason why I objected to unrestricted warfare was the inadequate number of our U-boats. As the struggle with the Press went on, the methods used became more and more unscrupulous. Political ignorance and malice culminated in the wicked suggestion that I was obstructing unrestricted submarine warfare only from pro-English or even from more disreputable motives."

We now come to the late autumn of 1916. Who was it who eventually did succeed in bringing about unrestricted submarine warfare and with it the entry of the United States into the war? Friend and foe alike exclaim: Ludendorff! Bethmann Hollweg says: "It was a change at General Headquarters in August 1916 that decided the question of submarine warfare". Churchill, who has made a careful study of the German war documents, decides that the responsibility undoubtedly rests with Hindenburg and Ludendorff. And Dr. Rosenberg, with the entire documents of the Parliamentary Committee at his disposal, comes to the conclusion that if Ludendorff, "the greatest soldier in German history", had, in 1916, been a lecturer at the military academy, or had been discussing the German position theoretically in the course of a *Kriegspiel* at Headquarters, he would have discussed the position of Germany in entirely different terms, and would have weighed the arguments for and against submarine warfare on their own merits. In 1916, however, his judgment was not that of a military expert but that of a politican who allowed his military judgment to be biased by a political illusion—the hope of a victorious peace.

The fatal decision was reached in the following manner. On the 31st of August 1916 a thorough discussion of the submarine problem took place at Pless in the presence of the Kaiser and the Chancellor. Hindenburg and Ludendorff gave it as their opinion that unrestricted submarine warfare was a suitable "and indeed the only means of bringing the war to a successful conclusion within a reasonable time", adding, however, that in view of the military position the moment had not yet come to have recourse to this weapon. "In forming an opinion ... of the value of submarine warfare", Ludendorff says, "we had to rely upon the judgment of the Admiralty and the Chancellor." If the distribution of power between the political and military leaders remained unchanged, the opinion of the Chancellor, provided he obtained the concurrence of the Kaiser, would even now have been paramount.

What was the position of Bethmann Hollweg? What information was available, and whence could he draw the power to oppose the advocates of the submarine; how was he to remain true to his convictions if they continued to conflict with those of the new Commanders-in-Chief? Bethmann Hollweg's position was weakened by the fact that the Reichstag had not been inactive. At the end of March 1916 the Central Parliamentary Committee adopted with one dissentient vote a resolution unequivocally recommending unrestricted submarine warfare. Conservatives, the Centre, and Social Democrats were agreed that "the time has come to make use of submarines as well as all other military means calculated to lead to a peace securing the future of Germany". On the 6th of April 1916 this motion was adopted by the House, and the resolution was communicated to the Chancellor in a letter signed by the President of the House.

On the 29th of September 1916 the Central Committee resumed its discussion of the submarine question. The Chancellor was present, and at that time still held a controlling position; in his view the only possible solution of the question was a temporary one. After a very brief interval, on the 16th of October 1916, the decision was made which caused the Chancellor to fall and gave the power to Ludendorff. It was the Centre Party whose action led up to this important event. The chief organ of this party, the *Kölnische Volkszeitung*, had been carrying on a campaign for unrestricted submarine warfare for months, and by the middle of October the entire parliamentary party had been converted to its views. A motion was brought before the House declaring that the Chancellor had sole responsibility for the conduct of the war in its political aspect; this declaration, however, was followed by the words: "The decisions of the Chancellor will have to depend largely upon those of General Headquarters. If unrestricted submarine warfare is decided upon, the Chancellor can rest assured of the approval of the Reichstag." This motion was adopted with a large majority, the Conservatives, the Free Conservatives, the National Liberals, and the Fortschrittspartei voting with the Centre. The meaning of the motion was clear enough: General Headquarters was to decide upon policy and the Chancellor was to bear the responsibility. Constitutionally the motion was absurd. It is true that the Chancellor was the only responsible person within the Reich and that he maintained touch between the Kaiser and the Federal Council as well as between the Kaiser and the Reichstag. But he was subordinate only to the Kaiser, who had an unrestricted choice in appointing the Chancellor. In every other way the Chancellor was independent. It may be added that the right of appointing the Chancellor was the civil privilege of the Kaiser, besides which he held supreme command of the army. Constitutionally the Kaiser was the Commander-in-Chief, and the Chief of the General Staff merely assisted him in a consultative capacity. The motion brought in by the Centre not only reduced the Chancellor to the position of figurehead, covering the acts of the Chief of Staff, but also abolished the supreme power possessed by the Kaiser as War Lord.

The motion adopted by the Reichstag reveals how thoroughly the distribution of power had changed by 1916. Under the influence of public approval, that of Ludendorff continues to grow; the Chancellor steadily loses support and the Kaiser has ceased to exist. Bethmann Hollweg, in his *Betrachtungen*, does not exaggerate when he says: "By the decision of the 7th of October 1916 the Reichstag surrendered political power to the military Commander." It was this motion which laid the foundation of Ludendorff's dictatorship, and very few steps remained to be taken before the alliance formed between the Reichstag and General Headquarters would be strong enough to supersede the Chancellor. It was an error due to Bethmann Hollweg's weakness that after the failure of the Centre he did not resign. Such a step

DICTATORSHIP

would have given him the opportunity to proclaim his opinion and to support it by a more vigorous propaganda. By such an act of intellectual honesty he would have enhanced his personal position in the eyes of history and, what is more, he would have rendered the greatest service to his country.

What followed now was no more than the execution of a measure that had already been resolved upon. At the same time the peculiarly German part of the tragedy, the tragedy of the experts, was only now about to commence. One expert appeals to the others, and none of them admits that he is blind and ignorant about the one essential part of the problem. Thus, Ludendorff writes: "The Chief of the Admiralty Staff, Admiral von Holtzendorff, an eager champion of unrestricted submarine warfare, positively declared that such a policy would suffice to decide the issue within six months". Thus the General relies upon the Admiral, and the Admiral in turn derives his wisdom from yet another expert. "He did not simply hold his view as a matter of routine, but also appealed to the opinion of leading German industrialists." It has been observed again and again that these representative industrialists, while possessing expert knowledge within their own sphere, where they are extremely exact and careful in their statements, become foolish blunderers once they go beyond their last. Their business successes, and the wealth and admiration they enjoy, lead them to adopt a note of authority when judging matters of which even captains of industry and successful merchants must necessarily be completely ignorant.

Among the mass of calculations and expert opinions which formed the material on which the resolve to enter on unrestricted submarine warfare was based, the Admiralty figures on the further construction of U-boats were approximately reliable; the estimates of their effect and of the antisubmarine measures of the enemy, and, worst of all, of the military power of the United States, were mere guess-work without foundation. Accordingly it may well be asked why Ludendorff allowed himself to be misled, and whence came the nervous impatience to enter upon unrestricted submarine warfare. It would be incorrect to assume that he was acting simply as a politician who allowed himself to be biased by his longing for a victorious peace, since he admits himself that the considerations which led him to seek safety in the suubmarine were of a different nature. His view of the military position was gloomy, and in unrestricted submarine warfare he simply saw "a last resource". A telegram of the 26th December 1916 from General Headquarters to Bethmann Hollweg describes the submarine as "the only means of carrying the war to a rapid conclusion. . . . The military position does not allow us to postpone this measure."

Ludendorff records that this telegram was drafted "under the influence of the impressions I had once more received in the middle of December during an inspection of the Western front". Thus Ludendorff's insistence on submarine warfare was not due to political but to military considerations: he had begun to fear that the war could not be won upon land, and he was still intent upon the military advantages of this measure. "Submarine warfare would improve the position of our army. Once the enemy begins to suffer from a shortage of pit-props, and a shortage of coal arises, they will be handicapped in the production of munitions. This would mean a relief of the Western front: the troops must be spared a second battle of the Somme." It is precisely because Ludendorff was not a politician that these calculations were faulty. He failed to see the new enemy; he did not perceive the enormous resources of the United States, and continued to believe that "certain concessions" in the manner of conducting submarine warfare would ensure the neutrality of the United States. But the submarine question was a political question, and it required a politician and not a soldier to solve it.

This was the position when, at a decisive session on the 9th of January 1917 at Pless, the only man who had preserved an independent opinion and a correct political instinct renounced his better judgment and capitulated before the Generals. Bethmann Hollweg's version of the position makes it appear that resistance had now become useless. "When I arrived at Pless early on the 9th of January 1917 the decision had already *de facto* been taken. General Headquarters and the Admiralty Staff had finally decided upon submarine warfare, and the Kaiser had sided with them. In the summer of 1916 I had succeeded in preventing submarine warfare, but since that time the situation had changed completely. At that time I had been able to enforce my view because the authority of General von Falkenhayn did not suffice ... to force through a measure which, while supported by the Conservatives, the National Liberals, and the Alldeutsche Party, was regarded with scepticism by the majority of the Reichstag." On the 9th of January 1917, however, the Chancellor continues, he was dealing with men who had made up their minds and refused to give way. ... "I declared myself incompetent to criticise the judgment of the military experts who insisted that the war could not be won on land alone. ... In view of these facts and of the declared readiness of Headquarters to risk war with the United States, I could not advise His Majesty to do other than to accept the opinion of his military advisers. Thereupon the decision was reached. ... On the next day", so Bethmann Hollweg concludes his melancholy tale, "General Headquarters asked for the immediate appointment of a new Chancellor."

Bethmann Hollweg's story is plain enough: the Chancellor gives way to the Generals—the political leaders to the War Office. The 9th of January 1917 is the real end of Imperial Germany. Once unrestricted submarine warfare had been resolved upon, the United States were bound to join the Allies and to place their full power at the disposal of the enemies of Germany. The final decision would not rest with the submarine at sea, but with the United States army on land.

CHAPTER IV

DICTATORSHIP (*continued*)

Submarine Warfare—Tanks—First Retreats—Failure of Submarines—
Fall of Bethmann Hollweg

MELANCHOLY as are the events of the 9th of January 1917, there is nothing to convince us that they were inevitable. It is true that Bethmann Hollweg was fighting an unequal battle, and that he, supported by a minority of cool and sensible people, had to combat General Headquarters, the Admiralty, the Reichstag, the Press, and three-quarters of the entire population. Nevertheless he might have won his case if the Chancellor and the Kaiser had known enough of German history to remember that the Prussian wars of 1866 and 1871 against an external enemy were accompanied by an internal strife between politicians and generals. It is the nature of a general, unless he happens to be a man of genius, to be without political understanding; and in 1866 Bismarck had to carry on a desperate struggle with the King and his generals before his own counsels prevailed. Again in 1870 a similar tension prevailed when Bismarck attempted to cut short the course of the war in order to avoid neutral intervention. And it must further be remembered that the General who opposed Bismarck was no less a figure than the elder Moltke.

Since Hermann Oncken has edited the Diary of Grand Duke Friederich of Baden during the Versailles negotiations we possess a brilliant account of the gigantic intellectual struggle which culminated in the victory of statesmanship over the generals, and with enabled the military victories to be exploited politically. When the siege of Paris in 1870

DICTATORSHIP

began to last longer than Moltke's calculations had foreseen, and the danger of a *levée en masse* became imminent, Bismarck grew apprehensive of intervention on the part of the neutral Powers. His demand that the military operations should be cut short amounted to an interference with General Headquarters, which led to an actual estrangement with Moltke and went so far that the Chancellor and the Chief of Staff refused to speak to each other; the King, to whom each came with his separate complaints, had to act as intermediary between them. On this occasion the statesman refused to capitulate. The struggle continued, for Bismarck was firmly resolved to avail himself of the first opportunity of entering into negotiations for ending the war. Moltke thought differently; his military mentality had only one aim —that of the unconditional surrender of the enemy. In the Grand Duke's Diaries there is recorded a discussion between the Crown Prince of Prussia (the father of William II.) and Moltke which is interesting if only because the Crown Prince's straightforwardness was of a very different quality from the various opinions which were voiced in connection with the submarine question of 1916.

The Crown Prince began by asking what Moltke intended to do once Paris had been captured. Moltke advocated the continuation of the war, with an advance southward in order finally to break the enemy's resistance. The conversation proceeded as follows: Crown Prince: "Supposing that we exhaust our own powers and are not always favoured by victory." Moltke: "It is up to us to see that we are victorious; and if we should happen to be beaten, we must rally our forces to win back the lost ground. We must fight this nation of liars to the last, until we have inflicted a decisive defeat upon them." (The reader will note that military language did not change between 1871 and 1916.) Crown Prince: "And what would happen then?" Moltke: "Then we can make our own terms of peace." Crown Prince: "And supposing we bleed ourselves white in the process?" Moltke: "That is an unlikely contingency, and even if it should come about, we shall at least have won peace." Crown Prince: "May I ask whether you are really acquainted with the political situation and think it sound politics to act as you suggest?" Moltke: "I have no view on these subjects. It is my business to consider the military aspect. If His Majesty gives different orders, I shall make different proposals accordingly." Crown Prince: "I would share your opinion if we were dealing with one or two Army Corps. The fact, however, is that we are dealing with the entire army; the war cannot be conducted from a purely military point of view, and consequently it is essential for you to remain permanently *en rapport* with the political views of His Majesty's Government and to act in conformity with them."

Moltke, however, remained obdurate, and continued to insist that Bismarck had no right to interfere in the conduct of the war. "It is no real concern", he records, "of Count Bismarck's; only he has the craving to have a voice and to give orders everywhere." Moltke's description of the facts is undoubtedly correct; and, indeed, Bismarck could not allow the General to act autocratically without endangering the success of the whole war. In the first instance, the Crown Prince attempted to smooth over the differences, and Bismarck and Moltke were invited to meet him to discuss matters. But this attempt at mediation proved a failure and it became necessary for the Kaiser to interfere. On the 25th of January 1871, Kaiser William, without consulting Moltke, issued two Cabinet instructions to the Chief of the General Staff by which the conflict was settled once and for all. It was ruled that all news of any importance must be communicated forthwith by the General Staff to the Chancellor; and further, the General Staff was forbidden to take steps of political importance without previously consulting the Minister for Foreign Affairs. Moltke was deeply incensed, and drew up a report, addressed to the Kaiser, in which he refused to admit the obligation to inform Bismarck of impending military operations unless the Kaiser made Bismarck responsible for the conduct of the war. "The military operations must be under a unified control, and I am perfectly ready ... to leave the sole responsibility to the Chancellor." This was merely another way of saying that if Bismarck was to interfere in military questions at all he had better undertake the entire conduct of operations. It is to the credit of Moltke that he did not despatch this memorandum. On the following day, when he took a calmer view of the situation, he composed another letter in which a statement of fact took the place of the ultimatum. In this version he declares that the Chief of Staff and the Chancellor are "two equal and independent officials under the direct orders of your Majesty, and it is their duty to keep each other informed. I on my part have complied with this duty, but it has not been done by the Chancellor." Once more, however, Moltke refrained from placing this despatch before the Kaiser. "With characteristic self-discipline he compelled himself to accept the Sovereign's decision in silence." The decision had, in any case, been taken already; on the previous day General Stoss, one of Moltke's assistants, had recorded in his diary: "Count Bismarck is negotiating with Jules Favre without taking any notice of General Moltke and the General Staff".

There could be no more instructive example than the above. It shows that the conflict between politicians and generals was not peculiar to the period of Ludendorff's ascendancy during the Great War. It had existed no less acutely during the Franco-German War, and many other instances could be produced to show that this conflict will always arise where the Commander-in-Chief does not also control foreign policy. The conflict of 1871 allows us to perceive what was missing in Germany forty-five years later. It would be incorrect to say that the Generals had the best of the political leaders in the dispute about Poland; the fact is that there was here no difference of opinion. And in the case of the second great decision—that about submarine warfare—it is necessary only to imagine that Ludendorff had been opposed by a statesman of equal prudence and calibre in order to obtain a correct view of the situation. If Bethmann Hollweg had been in Bismarck's position in 1871 he would have yielded as he did before Ludendorff in 1917; Moltke would have had his way, there would have been no negotiations with Jules Favre, and in all probability the war would have continued until the neutral Powers intervened.

This outline of the events of 1917 would be incomplete

without mention of the last chapter following the decision of the 9th of January. On the 29th of January there was a final consultation at Pless. The Kaiser had summoned the Chancellor, Dr. Zimmermann, Hindenburg, and Ludendorff. The question for discussion was President Wilson's attempt at peace mediation. The German conditions were outlined. Ludendorff's remarks are as follows : " The Chancellor never demanded postponement of unrestricted submarine warfare. . . . The discussion was rapid and took place in a private apartment of the Kaiser's. The birthday presents were still standing about. . . . I mentioned to Hindenburg that I deprecated the manner in which we had been asked to collaborate in such a vital decision. On the one hand we were not informed of all the facts, while on the other we had to share in the moral responsibility."

On the 31st of January 1917 the German note was handed over at Washington declaring the commencement of unrestricted submarine warfare.

When Ludendorff took over his new command, his aim had been to exploit every resource of the nation for war purposes. War supplies of every kind were to be increased. He was convinced that the legitimate demands of the army had not been adequately met, and he began by demanding conscription for the entire population, including women, between fifteen and sixty. The Act passed by the Reichstag for this purpose fell far short of his demands. The so-called Hindenburg programme drawn up by Colonel Bauer contained the demands made upon industry. This programme pursued much the same aim as had been enbodied at about the same time in England by the Ministry of Munitions.

The advantage of the Ministry of Munitions consisted in the fact that in its 50 departments and 12,000 officials the most capable personalities of English business life were united. Most of these men had given up their own undertakings and had made considerable financial sacrifice in order to place their knowledge and energy at the service of the State. They received no salary; industrialists and men of business served entirely from patriotic motives. These men were assisted by a staff of permanent officials. The various

DICTATORSHIP

departments of the Ministry were kept distinct; important questions, however, were settled jointly by the committees. This combination of business experience and official routine proved very successful, and it is no accident that one of the most important military inventions was appreciated at its proper value by this body of men of business. When the first tanks were introduced, the military experts who witnessed the initial driving experiments were very sceptical, and Mr. Churchill, the chief supporter of this invention, was compelled to carry on deductive propaganda before he was enabled to demonstrate their value practically. It was once the pride of the Prussian General Staff to point to the logical methods of Clausewitz as an inalienable heritage. In this instance it was the English who were led by logical processes to recognise the necessity of finding a new engine of war capable of effecting a saving of man-power.

Mr. Churchill drew up two memoranda dealing with modern trench warfare, in which he showed how it still made frontal attacks inevitable, although this method had been abandoned by theorists forty years ago. In the absence of any open flank, the only possibility of defeating the enemy consisted in a break-through. The defence, however, was now so strong that it was not sufficient for the attacking to be superior to the defending force. All the offensives in the West taught the same lesson: that the greatest courage and sacrifice on the part of the attacking force did not possess sufficient mobility and destructive power to break down the defence. A maximum concentration of artillery on any given sector could guarantee no more than the destruction of so many square miles of battlefield and a strictly limited advance. The artillery was of merely local effectiveness; its destructive force was enormous, its mobility almost nil. The only branch capable of movement was the infantry; and the infantry had not the power to overcome the resources of a modern defence or to carry an offensive operation to a strategic success. A machine had to be invented capable of taking the place of human beings under the severest conditions before even a superior force would be able to make a steady advance. These considerations, based upon the conditions of modern battle and the limits of human endurance, fell on fruitful soil. It was not a body of military experts, but a number of English business men who provided the Allied armies with a weapon which played a paramount part in the concluding battles of the Great War.

In agitating for an independent central department to control the conduct of the war in all its economic aspects Ludendorff was inspired by the correct idea. He only succeeded, however, in having a War Department, consisting of three sections, set up within the War Office; he and Colonel Bauer were the controlling brains, while the War Department was charged with the realisation of the programme. Ludendorff tells us that the industrialists played their full part in the war, and that this would always be to their credit. "In view of the risks and the great financial sacrifices, they had a right to demand adequate remuneration from the State, no less than the workers had a right to demand good wages."

The first tank attack was delivered during the battle of the Somme in September 1916. The Germans were completely taken by surprise. Churchill would have preferred to keep this new arm, which had intentionally been given an unintelligible name, in reserve until a larger number of tanks should be available; he was prevented by the Generals, who, after first under-estimating their value, were now clamouring to have them brought into action. A certain time, however, had to elapse before the first models could be improved upon and the new arm could be rendered really successful.

Ludendorff's attitude towards the tanks shows an uncertainty which can frequently be observed in him. On the one hand, he looks at things in an old-fashioned manner, and produces platitudes like: "the best weapons against the tanks are coolness, discipline, and courage", as though there were anything heroic in having a duel with a locomotive. On the other hand, he recognises that such an invention must be countered by similar means. In October 1916 the War Office received instructions to study the construction of tanks. "Without over-estimating their value, one cannot deny that they have met with some success. In any case, an improved model would be an effective weapon. Accordingly I con-

DICTATORSHIP

sider it desirable that the construction of tanks be entered upon forthwith."

With the commencement of construction, mistakes began to be made. The factories were working at full pressure to turn out lorries for the transport of troops, which were urgently needed; accordingly the War Office attempted a compromise, and caused a model to be constructed which was to serve alternatively as tank and for the conveyance of troops. The actual construction was delayed by the military bureaucracy and by numberless commissions and inspections. Eventually the first German tank was ready for trials in May 1917; it did not really satisfy anybody. Thereafter the construction of tanks was taken over by the Mechanical Transport Department. By the beginning of 1918 only five tanks were ready, and in the course of the year, while enemy tanks were numbered by the thousand, the Germans made use of no more than fifteen German and seventy-five captured enemy tanks. Colonel Bauer had caused a light tank to be constructed by Krupps on his own initiative; when, in the spring of 1918, the French light tanks built by Renault first appeared, recourse was had to Bauer's model and the order was given for the construction of light tanks. But now it was too late.

The problem of the tanks was no more than a drop in the sea of troubles, of plans and designs which the year 1917 brought with it. Ludendorff looked with dismay at the maps with their front lines winding across them in endless curves. In the East winter imposed inactivity on the enemy, and no important movement was to be expected before April; in the West, however, everything indicated an impending storm. The opposing Generals could not see each other's cards; they were forced to draw their conclusions from the reports of airmen, the statements of prisoners, and other insignificant clues. Ludendorff was not in a position to guess Joffre's intentions, and could not know that a grand attack on the Somme and near Bapaume was due for the 1st of February 1917. He did not know what was going on behind the enemy's screen of arms; all that he could learn from the enemy press was that the hostile Commander-in-Chief had been removed and that a new man had taken his place. General Nivelle, the victor of Verdun, was the hope of France. He succeeded in winning the confidence of Lloyd George, and indeed his assurance and certainty of success were such as to dazzle all with whom he came into contact. "It does not matter", Lloyd George said, "that one general is better than another, but that he is better than two others", sanctioning with these words a kind of supreme control extending over the British Expeditionary Force. Nivelle was younger than Ludendorff, and his plans were of a wider scope than those of "poor Joffre". He also had novel tactical ideas—he did not desire a battle on a wide front; the assault was to be brief but of the utmost intensity. The German front was to be smashed at one blow of a gigantic fist; "brutal" and "quick" were Nivelle's favourite words.

The plan was still being worked out when an event happened which took the enemy Command by surprise. On the 24th of February the English observed the German artillery firing on their own trenches. British patrols found that they had been evacuated. During the following nights the horizon was lit by vast conflagrations, while enormous clouds of smoke rose into the sky all day. Ludendorff had withdrawn the German salient between Arras and Soissons and had taken up the Siegfried line twenty miles further back. Elaborate plans had been made to devastate a strip of land some ten miles wide before the new front.

"The great military personality which Germany had discovered in her need, by one sure stroke overturned all the strategy of General Nivelle. Towards the end of February the German evacuation of the whole sector from Arras to Noyon began. Leaving a screen of troops to occupy the abandoned positions and fire off their guns and rifles, the German army withdrew fifty miles from the threatened area of the salient, and with unhurried deliberation assumed their new deeply-considered positions on what was henceforward to be known as the Hindenburg line. The German General Staff called this long-prepared operation by the code name *Alberich*, after the malicious dwarf of the Nibelungen legend." This is Churchill's version of these events.

The enemies had a higher opinion of this step than Ludendorff himself, who admits that he had great difficulty in resolving on it. "It contained a kind of admission of weakness, which was bound to improve the enemy *morale* and to depress our own." In spite of the jubilation of the Press, little satisfaction was felt by the enemy. Joffre had intended to cut off this salient, which protruded from the German line with a front of 90 miles; Nivelle wanted to do the same, and other things besides. To cut off this salient would have cost at least 20,000 dead and 80,000 wounded; now the salient had vanished without any loss to the French or English forces. Nevertheless, Ludendorff's voluntary retreat was appreciated as a skilful move by the enemy Commanders—a tacit admission, incidentally, of the slight value attaching to a gain in ground, a truth not otherwise readily admitted. It is the sacrifices which such gains generally involve which compel the generals to describe them as successful operations worth the price. In fact, every offensive is a waste of men if it succeeds only in advancing the trenches a few miles. French and Germans alike were destined to learn this lesson on several future occasions.

The chief merit of the *Alberich* movement consisted in the intuitive certainty with which the region of the contemplated attack was devastated and evacuated. The main body of the attacking army was faced by country where an advance was almost impossible: everything, railways, roads, villages and wells, had been destroyed. Ludendorff records, however, that orders had been given that the wells must not be poisoned. Works of art within the region that was to be devastated were removed into occupied territory. "It was deeply regrettable that much property was destroyed; but it was inevitable." Ludendorff had been successful beyond his own intentions. He had desired to shorten his front, and in fact he did shorten it by 20 miles, which allowed him to withdraw an entire army and to utilise it elsewhere. But not only this; he also gained time to prepare against the impending assault, while Nivelle for his part had to postpone his attack and consequently to alter his plans. The battle had been intended for February; it actually began on the 8th of April with the British attack north and south of Arras. Later, the offensive was carried on by the French, this phase of the battle lasting from the 16th of April until the 9th of May. Nivelle's plans were ambitious; while the British were pressing on towards the east at Arras, the French were to effect a break-through on either side of Rheims, and the whole German front, as far as the sea, was to be shattered. This, the strategic aim of the battle, was not reached. Ground was gained, prisoners were taken by the French and the British armies, guns were captured; but although the German front was pushed back, it was not broken.

Ludendorff watched the gigantic conflict with the gravest anxiety. After the partial successes of the British forces, he records: "I had looked forward to the attack with confidence; now I was deeply depressed. Was this to be the result of all the troubles of the last six months? Were my tactical instructions wrong?" They were not wrong. The new tactics, with their loose front line, their elastic manœuvring, and the so-called emergency divisions which were used at points where danger threatened, had all proved their value. But it was now seen on both sides that the laymen of the British Ministry of Munitions were correct in their estimate of the limitation of human endurance. The German infantry continued to do miracles in defence, and where they were forced to give way, it was a sure sign that the mechanical superiority lay with the other side. French and English alike, advancing in thick waves according to the instructions worked out by Nivelle and Haig, failed, in spite of the utmost gallantry, to break through the German front. Here, too, the limit of human endurance was reached; but the destructive power of the defence was great enough to bring infantry movements to a standstill. The theory of the brief and brutal blow overlooked the fact that, after the first stages of the attack, the parts are changed: the attacker was forced into the defensive, and the defending side began to attack. While the artillery of the attacker was following up laboriously, the attacking infantry was exposed without support to the concentrated storm of fire let loose by the defence. It is at this moment that the fate of an offensive is decided.

Ludendorff consistently checked the efficiency of his forces. He distinguished between war-weary and devoted divisions, and, with characteristically elementary psychology, measured all the human emotions on the field of battle by the simple standard of discipline. It would be too much to say that this crude method is peculiarly Prussian; it is to be found with most generals, and among the French and British no less than among the Germans. Orders and addresses to the troops are the same in every army. With courage they have very little to do; and, after the most extravagant of all wars, it is surely inadmissible to apply a measuring rod to the physical courage of modern fighters, still more to claim that a certain amount of courage is sufficient to decide the event. What does decide as between a general and his forces, and, indeed, between every leader and the men he leads, is the confidence that is felt. Mutinies such as occurred in certain units of the Austrian army served only to indicate that there was no confidence between soldiers and general, between citizens in uniform and the State. Where there is no will to fight, orders, punishments, and the strictest discipline are unavailing; and where the will to fight was manifested in its greatest intensity, as it was in the armies of the Great Powers, every decrease in the will to attack was a sure sign that confidence was on the wane.

It is not the case that death is the same, whether it comes in a bayonet charge or a cavalry attack, whether it advances behind a tank or destroys thousands in clouds of poison gas. A brave man clings to the illusion that when he is compelled to stake his life, some element of personal struggle remains, in spite of every development in the technique of destruction. Self-sacrifice and death only appear meaningless once he grasps that there is no relation between his own effort and the destructive power of the enemy, and when he sees that his courage and endurance are wasted. When the battle is carried on essentially by mechanical means, the soldier feels the maximum tactical and technical protection to be no more than his due. And modern battles of destruction have shown that such protection is not afforded by orders to preserve discipline or to hold out to the last man. "The only end can be the destruction of those who hold out", Hans von Hentig remarks: "The experienced soldier sees in it a sign of the inadequacy of their commanders and of their inability to discover better methods; he feels that he is sacrificed in vain, and that the leaders are shifting their responsibility upon himself. Where careful and effective measures are taken for the safety of the men, such care is repaid by the utmost devotion; but sooner or later troops will resist, if only by passive methods, useless destruction by shells and gas. They punish a stupid or incompetent leader by allowing themselves to be defeated, and in their desperate position have recourse to any principle that will justify such conduct. A general who must be defeated before he learns that his men have lost their mettle never understood his business."

While the battle of Rheims and the attacks on the Chemin des Dames were in progress, a dangerous mutiny broke out in the French army. At first a single division refused to advance against the hail of German shells and bullets. Presently, however, the mutiny extended, and eventually extended to 16 Army Corps. A number of regiments turned about, marched back to Paris, and refused to fight under generals who wantonly sacrificed their men. The mutineers were answered by machine guns; and there was no man in authority with sufficient psychological understanding to appreciate this protest of men who knew that their lives were being wasted. A veil of profound secrecy was drawn over the graves of those who were court-martialled. But the sacrifice was not in vain. The French Government took the only step possible in the circumstances; Nivelle was removed from his post and Pétain became Commander-in-Chief.

While the French army was shaken by the failing confidence in its leaders, similar doubts began to arise in Germany. The kind of distrust that was felt was of a peculiarly German nature. It was not directed against General Headquarters, nor against Ludendorff, whose authority had been diminished by no reverse and accordingly remained unchallenged. Nor did it arise from general considerations; nobody arose to ask what were the aims and meaning of the war. In matters like this, logical thought was subordinated to a vague sense of superiority, based upon the unshakable belief in the invincibility of the German soldier. Those who feared that a war without annexations or indemnities might diminish their power or endanger their interest constituted a minority; the majority were never interested in the objects of the war, and were content to allow the demand for an acceptable peace to hide its materialistic nature under a nationalistic or metaphysical cloak. It was not only the leaders of German heavy industry nor the supporters of the old régime who based their claims for a greater and more powerful Germany upon the victories and conquests of the German army.

A large part of intellectual Germany—scholars, poets, and the press—were intoxicated by this idea of a "strong peace". Rudolf Borchardt, a magician of language, succeeded in charming industrialists and bankers, sober men of figures and thinkers alike, with his image of a victorious peace. Vague notions possessed many minds, to which he gave a form in words when he explained that the Great War was the will of God, a kind of fiery furnace to test the peculiar gifts of Germany, and to impose a reasonable order on the world once Latin Civilisation had been brought to the ground. One of the wicked Germans, Karl Marx, on one occasion said that the difference between Germans and English was that a German would change a hat into an idea, and an Englishman an idea into a hat. But besides this, which Oswald Splengler calls the Faustian tendency, there exists in every German a belief whose proper sphere is not a Faustian heaven but the sober German workroom: the belief in the expert. Any German allows himself to be led by the tunes of Faustian ratcatchers; charmed by their notes, he will follow them along the most dangerous paths. But on the other hand, he will also continue to believe in the expert—until he is disillusioned. The belief in Ludendorff remained unshaken because Ludendorff had never lost a battle.

But with the submarine campaign doubts began. Technically the submarines did not prove a failure; indeed, their success was considerable for their numbers. But the promises of the naval experts were demonstrated to be illusory; the latter had promised that within six months England would be paralysed, and they had asserted that the submarines would act with mathematical certainty. Hans von Hentig inclines to think that it was a delight in merely material weapons which allowed the Germans to over-estimate the value of a torpedo moving invisibly through the sea, of a mine lurking in the waters, of a long-range gun bombarding Paris, or of a Zeppelin suddenly emerging over London. "We aroused the hatred of the whole world, which only those have a right to challenge who are strong enough to beat it down." In fact, however, Germany was no "modern Macedonia" threatening civilisation; the motive power was the German trust in exact science, theoretical and applied, acquired in school, laboratory, workshop, and factory. The technical expert was the priest of this belief; should he fail, his doom was sealed.

A long road had to be covered between the first movement of distrust in the submarine and definite scepticism. Waiting millions were calculating and checking figures, until the promised success of this "last measure" should become apparent. Ludendorff misunderstands his countrymen if, looking back upon these days, he speaks of a growing spirit of defeatism. "It became clear to me at that time", he says, "how far we had already drifted. If things were to go on in this way, and nothing was done to strengthen the morale of the nation, a defeat was inevitable." Again, in a letter from General Headquarters to the Kaiser, he blames Bethmann Hollweg for the declining spirit of self-sacrifice and aggression noticeable among the population—as though the Chancellor had missed a single opportunity of fighting the decay in confidence by popular addresses designed to create a patriotic enthusiasm. It is a curious phenomenon that Ludendorff, the military expert, faced by the failure of the naval experts, has recourse to what he imagines to be ethical arguments. The failure of confidence of the summer of 1917 had concrete causes: the populace, which had been trusting its experts with superstitious loyalty, for the first time saw its faith deceived. It was this doubt which called up contradiction, awakened opposition, and found a ready voice even in the Reichstag. Naturally enough, the true political aspect of Germany was revealed by this failure of confidence.

The loss of confidence in the war leaders had more lasting effects than the repercussions of the Russian March revolution. The greatest event of modern history, the fall of the throne of the Czars, failed to fructify political thought or to call forth any constructive criticism. The Left wing Socialists were encouraged, and the split within the Social Democrat party was hastened; but it called forth no statesmanlike ideas. And indeed, for the moment it was impossible to forecast developments in revolutionary Russia. The bourgeois democracy in Russia was unwilling to break off the war; and it would have required an expert knowledge of Russian psychology to foresee that every appeal to arms must fail once the flag of peace is unfolded. Meanwhile the men of to-morrow were in exile in Switzerland. Ludendorff was well advised when he recommended that the Bolshevik leader Lenin should be allowed to travel across Germany; but the Russian comet passed across the country and left no trace. It is not the function of a general to have political imagination; unfortunately the German politicians, too, were lacking in this essential. Ludendorff saw in Lenin only the destroyer who would succeed in finally demolishing the Russian power of resistance. "The people and the army alike were rotten, otherwise the Revolution would have been impossible. . . . Often and often I had hoped that a Russian Revolution would come to ease the military situation; but every hope had been a castle in the air. Now that the Revolution had come, I was taken by surprise. A weight was off my mind. It was impossible at that time to foresee that the Russian Revolution would one day undermine the strength of Germany."

The growing distrust of and opposition against the German war leaders were not a result of the Russian Revolution. These revulsions were due to the disappointment of the hopes raised in submarine warfare. The Admiralty Memorandum of the 7th January 1916 had stated that it was "practically certain" that unrestricted submarine warfare "would break British resistance in six months at the outside". Admiral von Holtzendorff had said: "I give my word as an officer that not one American will land on the Continent". On the 9th of January 1917 unrestricted submarine warfare commenced.

Obediently and with complete trust in the opinion of experts Germany waited for the six months to pass. Almost to the very day when this term expired, criticism commenced.

In Parliament it was Herr Erzberger, a member of the Centre Party, who was the spokesman of the reawakened suspicions. His great speech on the 6th of July, delivered in the Central Committee of the Reichstag, was the first blast of distrust. Erzberger was well aware that any criticism must fasten upon the submarine campaign; this knowledge was the German element of his nature. But the wily politician, whom many threads bound to the great world of Catholicism, knew far more than he said. He received information from the Vatican and from Count Czernin; he knew the weakness ot Austria and the readiness of the Habsburgs to desert Germany, and with the help of his informants he was able to form a better view of the state of Europe than all the experts in Germany. In his memorable speech he raised the question whether, by next year (1918), Germany was likely to obtain a better peace than she might now by giving up her claims on the occupied regions. This question he found himself unable to answer in the affirmative. While he was of the opinion that the front line would hold, he feared that the fighting apparatus of the enemy would grow too fast for Germany to keep up with it, even with the utmost economy.

The hopes placed in the submarine had proved illusory, and Erzberger produced figures which made his arguments irrefutable. The calculations of the Admiralty were wrong. Erzberger pointed out that if it was desired to study the position in other countries, it was a mistake to rely upon Press reports, since the Press was under censorship everywhere. It would be better to read the advertisements. In Germany substitutes were being advertised; in England raw materials of pre-war quality were being sold, while in Paris butter was being offered at 18 per cent above pre-war prices. It would be wrong, Erzberger continued, to allow such considerations as that the war must now be allowed to have been waged for nothing, to carry weight, since by next year the war debt would have increased by some fifty milliards of marks. A continuation of the war simply spelt ruin, and it would be necessary to go back to the idea of defence which had prevailed in the beginning. While the High Command would have to continue working at full pressure, as hitherto, it would be essential for a large majority to form in the Reichstag, unequivocally admitting the idea of a defensive war as it had been laid down on the 1st of August 1914. If the German Reichstag made a gesture of this kind it could not be interpreted as weakness, especially if it were made clear that Germany would fight to the last man if the overture were rejected. Concluding, Erzberger pointed out that it was only Bismarck's moderation which had led the military operations of Germany to a victorious conclusion, and that, while the German army was aiming at a maximum, the political programme of Bismarck had aimed at a minimum. Contrary to the desire of the generals, he had in 1870 made three attempts to bring about an armistice with France, nor had this prudent moderation ever been interpreted as weakness. It was the duty of the Reichstag to conduct German policy in this sense, and it must not be allowed to expose itself to the heavy charge of having seen the true course too late.

An objective historian will admit the purity of Erzberger's motives and the rightness of his political judgments; nevertheless, such students will have to charge him with two mistakes. It will be urged that he criticised Bethmann Hollweg when he should have been criticising Ludendorff, and that, when his speech had led to the formation of a parliamentary majority, he did not go on to form a parliamentary Government. The first charge is baseless; it was politically correct to appeal to the Chancellor, since the Chancellor carried the full responsibility and controlled the general policy of the war. Only if Bethmann Hollweg had proved incapable of carrying out this task would it have been right to look for a substitute. The second charge is valid. The Kaiser had placed himself in the background, and no constructive policy or energetic action was to be looked for from him; but Bethmann Hollweg, in spite of many good qualities, was not the man to stick to a course which he had found to be right and to transform it into action. Once again a position had arisen where only a single politician appreciated events correctly and pointed out the right course. This man was von Richthofen. He was aware of the lack of programme, of the renunciation of the Kaiser, and of the inadequacy of the Chancellor. To the pressure exerted by General Headquarters no capable politician stood opposed. Yet it was necessary to find a programme, and to find a man to take up the most important post in the control of war policy; nor would it have taken a vast amount of courage to take the necessary step. But the Reichstag alone could take this step. von Richthofen urged Erzberger in this direction, and besought the majority parties of the Reichstag, now that they had been galvanised into action, to take control.

At this moment, when a correct appreciation had at last been formed, the scene suddenly grew dark and a tragedy of errors and confusions commenced. Three days after Erzberger's speech, Stresemann spoke. He supported Erzberger in all his attacks upon the weakness and vagueness of the Government; he criticised their lack of resolution, and claimed that they failed to maintain a warlike spirit in the nation. Further, he demanded a change in the Prussian franchise, and regretted that the Kaiser was "too completely mummified" to realise the ideal of a Kaiser who should be a truly national leader. Going further than the Centre and the Social Democrats, he went on to demand a parliamentary Constitution for Germany. The whole tenor of his speech contained an undisguised and urgent demand that Bethmann Hollweg should be replaced. "The whole conduct of the nation's affairs", he exclaimed, "is being carried on under the motto 'We shall not succeed anyhow'. Essentially the prevalent defeatism is due to the fact that the nation believes that it is moving from failure to failure in this greatest of all wars. . . . This tension is more than the nation can bear in its present position. A political defeat of the utmost gravity is inevitable." He would not admit the excuse that the politicians' hand was being forced by the generals. "A Chancellor must succeed in having his way; if he fails, he must draw the necessary conclusions." It was the first time in German history that a member of the Reichstag demanded a change of Government in such tones. A part of Stresemann's speech was on all fours with that of Erzberger; in the main, however, its tenor was quite different. Stresemann in his rôle as parliamentary agent of Ludendorff asked for Bethmann Hollweg's head. What frightened Erzberger was the vagueness of a programme which substituted the idea of holding out to the bitter end for any political initiative. What he considered most important was a return to the firm ground of defence from the vague mist of annexationist plans: he wanted to see the war ended before it was too late. Stresemann, on the other hand, felt confident of victory and made no secret of his annexationist ideals. This admission he put in parliamentary language by saying that it would be a piece of characteristically German folly to surrender the advantages of the military position and the possibility of dictating peace, from any theoretical reasons. These are Ludendorff's ideas clothed in the vigorous rhetoric of Stresemann. Whatever opinion may be held about these two speeches, and whether Erzberger is held to be the sounder judge of the international position, or, on the other hand, Stresemann is excused on the ground of his more violent nationalistic impulses, it is certain that neither saw in Ludendorff a source of danger. What both demanded was a real statesman, a man with a positive political programme.

At the same time the Crown Prince came on the scene. His campaign was inspired by Colonel Bauer, who had published a pamphlet as early as March 1917 in which he urged the necessity of a new system. In it Bauer complains that many of his comrades called him a semi-Socialist, and that he was looked at with doubtful eyes by those "honest and loyal men who, later on, were the first to declare their allegiance to the republic". On the other hand, he mentions that in Government and Trades Union circles he was treated as a reactionary and an agent of the capitalists. An unbiased judge will see in his well-meant memorandum a primitive kind of Fascism of the Hitler type.

Here we find yet another melancholy instance of the tragedy of the expert. This extremely able Prussian officer, whose technical and organising talents were quite exceptional, shows a startling ignorance and blindness outside his proper sphere. It makes painful reading to follow his argument about State and Society, politics and economics, in a vocabulary of which even National Socialist bravos would be ashamed. As Colonel Bauer himself records, he telephoned to the Crown Prince that he must come to Berlin. "I explained the position to him and convinced him of its gravity. The Crown Prince urged Bethmann's removal upon the Kaiser, but was refused. He declined to give way, and on my advice received most of the party leaders, with whom I had already opened negotiations so far as was necessary. It was found that the majority wanted to get rid of the Chancellor. It was Erzberger's opinion which decided the issue."

Details about the meeting called by the Crown Prince on the 12th of July have been supplied by Friedrich von Payer, the leader of the Advanced People's Party. It was the first occasion in forty-five years that a member of the ruling house of Prussia considered it necessary to inform himself officially of the views of the party leaders. Clearly the Crown Prince was less concerned to form an objective opinion than to achieve the fall of Bethmann Hollweg. Minutes of the meeting were kept by Colonel Bauer. The Crown Prince's questions were curt. Was it or was it not necessary to have a new Chancellor? To Erzberger he said: "Why don't you throw him out of the Reichstag?" The members under examination stood to attention with the next man by their side, so as not to waste time. In this manner Erzberger, Count Westarp, Payer, Stresemann, David, and Martin were cross-examined. The manner in which this inquisition was carried on was anything but edifying; the victims, however, were Monarchists and consequently accustomed to look upon the Crown as something of a higher order.

The manner of the conversation was less important than its matter, and on the latter point all were unanimous, with the exception of David. The Social Democrats, who throughout the crisis displayed a lack of energy relieved by no spark of imagination, found themselves unable to come to a resolve. They were not in sympathy with Bethmann Hollweg; but, since the Chancellor had succeeded in persuading the Kaiser to approve of an equal franchise for Prussia, there was no excuse for animosity. They were further convinced that they would succeed in winning Bethmann Hollweg's approval for the peace resolution which they intended to bring in. All the other parties were opposed to the Chancellor. The Centre voiced its views in a resolution to the effect that Bethmann Hollweg's continuance in office was an obstacle to peace, though they left it to the Chancellor to determine the moment of his resignation. The National Liberals informed the Chancellor's deputy, Dr. Helfferich, that in their opinion the crisis was unsurmountable unless the Chancellor resigned. The Conservatives were hostile to Bethmann Hollweg on principle. Thus the Chancellor's position was untenable even without Ludendorff's intervention.

The next act of the drama took place in the park of the Imperial castle of Bellevue on the 12th of July 1917. In the course of the afternoon the Crown Prince called upon the Kaiser. He reported his examination of the party leaders and his visit to Prince Gottfried Hohenlohe and to M. Rizoff, the Bulgarian Minister. The conversation turned upon Bethmann Hollweg, who had been commanded to appear at 7 P.M. On his arrival, the Kaiser received him ungraciously. He pointed out that he had expected Bethmann Hollweg to overcome the crisis, and that it was in this expectation that he had sanctioned the change in the Prussian franchise. The Chancellor objected that the reform was overdue in any case, that he had not urged it on his own account, and that equal franchise was the only conceivable franchise at the moment. The Kaiser agreed reluctantly. He next enquired about the peace resolution, mentioning that the Crown Prince had complained of its "flabbiness", and insisting that Hindenburg must be heard. The Kaiser himself was not yet acquainted with the resolution; Bethmann Hollweg drew the draft from his pocket and read it out. The Kaiser listened in silence; when the Chancellor had finished, he gave orders for the document to be telephoned to Hindenburg without delay. Within half an hour the reply from General Headquarters came to hand. It complained that the resolution omitted to thank the troops, and demanded two other alterations. The Kaiser agreed and requested Bethman Hollweg to inform the party leaders.

At this moment General von Lyncker, the Chief of the Imperial Military Cabinet, entered the room and reported that a message had been received by telephone from Kreuznach to the effect that Hindenburg and Ludendorff had handed in their resignations, and that the resignation of the whole of General Headquarters was on the way. The reason given for this step was that the two Generals found themselves unable to co-operate with Bethmann Hollweg. It was further stated that Ludendorff was resolved on this occasion not to give way under any circumstances whatever. The Kaiser was furious: he declined to yield to the ultimatum of the Generals, and caused telephonic instructions to be given them to proceed to Berlin forthwith.

Bethmann Hollweg knew that this was the end. He declared that it was naturally unthinkable that the two able Generals who enjoyed the confidence of the nation should be allowed to resign, and took his leave of the Kaiser. Next morning he tendered his resignation, at the same time suggesting Count Hertling as his successor. In order to allow the Kaiser to sanction his withdrawal without reference to the ultimatum of the Generals, the letter of resignation made no reference to the latter, and described the parliamentary situation as the reason for resigning. On the Generals' arrival, the Kaiser informed them that Bethmann Hollweg had tendered his resignation and that it had been accepted. They thus had no occasion to refer to their own intended resignation.

CHAPTER V

LUDENDORFF AND THE POLITICIANS

Failure of the Politicians—The Danger of the Expert—Michaelis and the Peace Resolution—The Campaign of 1917—The Pope's Peace Appeal—First Success of the Tanks

THE Chancellor had gone. But it was only now that the real tragedy was about to commence. All that had happened hitherto was that the forces of the Left and of the Right agreed in their opinion of the Chancellor and jointly urged that Bethmann Hollweg must go. Their motives were different. The one wing held that all military successes were useless unless they were exploited politically to reach a peace by negotiation; the other wing wished to carry on the war, and required a Chancellor of greater initiative and energy in order to obtain a victorious peace. If Dr. Bredt in the expert opinion submitted to the committee of investigation compares the action of the Reichstag majority to Bismarck's procedure in 1866, when the Chancellor made overtures for an equitable peace with Austria over the head of the King and the victorious generals, this comparison contains at least half the truth.

At the moment, however, a leader was wanting, capable of transforming military into political successes. The Reichstag majority undertook this task; but desire outran performance, and when it came to the actual transformation the Reichstag majority failed. The reason is simple. The majority could have enforced their wishes only if the new Chancellor and his Government were in agreement with them; and at this crucial moment the Reichstag remained inactive. Who was to appoint Bethmann Hollweg's successor? Constitutionally the right of appointment rested with the Kaiser: actually William II. would have been only too glad to pass on the responsibility. It had been with reluctance that he allowed Bethmann Hollweg to go, and he declined to accept Count Hertling, who had been suggested by the Chancellor on his departure. Nothing could be more characteristic of the weakness of this stage of Hohenzollern Germany than this crisis; the fate of the entire nation depended upon the choice of a Chancellor, and everyone was groping in the dark. The Kaiser and his advisers had no better resort than to entrust the Court officials, and more especially Count von Valentini, the head of the Civil Cabinet, with the task of finding a successor. The Kaiser fully appreciated the importance of the decision; but he felt too uncertain of himself to act upon his own choice. One name was examined after the other. For a moment Prince Bülow was considered, and Hindenburg himself, more courageous than most of the politicians, himself brought forward his name; the Civil Cabinet, however, had not the pluck to consider a man whose name had been on the black list since the *Daily Telegraph* scandal. On the afternoon of the 13th of July, Herr von Valentini suggested, "with a certain amount of enthusiasm", as Bethmann Hollweg records, the name of Count Bernstorff, the former Ambassador to Washington; but by 7 P.M. his claims, too, had been rejected. The Kaiser had declined to consider him further; Dr. Michaelis, the Food Controller, had been suggested, had accepted the call, and was already on his way to see the Kaiser.

The following events had led up to this step. After Count Bernstorff had been rejected, bewilderment prevailed for a while. The Kaiser was silent and, declining to suggest any name, instructed Valentini to ask Hindenburg whom General Headquarters wanted. Valentini, knowing his unpopularity at General Headquarters, hesitated before this delicate task. Eventually he approached General von Lyncker, with the sanction of the Kaiser, and acquainted him with his mission. During the subsequent interview, he met General von Plessen. The latter remembered having heard of Dr. Michaelis as a man well thought of at General Headquarters: he had visited Kreuznach, and had given the impression of energy;

LUDENDORFF AND THE POLITICIANS

he was, in military language, the kind of man "who got things done". Thereupon Lyncker and Plessen acquainted Hindenburg and Ludendorff with their discovery. Ludendorff agreed, and it was in this manner that the Food Controller, whose name was completely unknown in political circles, became Chancellor of the German Empire. When the Kaiser was acquainted with the choice, he had to admit that he had never heard of the man.

Even the enemies of Ludendorff cannot fairly claim that he brought about the fall of Bethmann Hollweg in order to place a nonentity like Michaelis in his place. It was not the General's intention to play the part of major-domo of the Hohenzollerns; still less was it his ambition to make a man like Michaelis head of the Government. But neither he nor Hindenburg held a candidate in readiness, and Ludendorff admits that he was extremely surprised when he found "that a substitute for the Chancellor was not permanently held in readiness by the competent authorities, and that in this vital matter Germany was living from hand to mouth". His surprise is characteristic of the military man. In the army a substitute was available for every officer from the Commander-in-Chief downwards, and it seemed incomprehensible that the same should not be the case in the civil administration. This typically military view is evidence in itself that Ludendorff had no desire to decide the selection of the Chancellor. He would have accepted any choice made by the Kaiser, even if the Chancellor so appointed had been a member of the Reichstag. The latter body, however, never made any attempt to play a part in the appointment. If we consider the preliminary sound and fury of the majority parties and the flood of rhetoric in favour of a parliamentary Government, and compare it with the subsequent complete collapse, we must agree with Dr. Bredt when he describes the appointment of Michaelis as one of the greatest muddles of history. After the fall of Bethmann Hollweg it lay with the Reichstag to seize the political control of Germany. The office was vacant. The Kaiser had ceased to act the autocrat. Ludendorff had no desire after dictatorship. The spectre of an autocratic and militaristic Germany would have been laid at one blow if the Reichstag was willing to act as common sense dictated. But the unique opportunity was allowed to pass.

The best comment on these events is that of Dr. Arthur Rosenberg, the Socialist deputy, who defined Ludendorff's destiny in the following terms. Ludendorff, he says, did not strive to be ruler of Germany. In August 1916, when army and nation clamoured for him, he did his work without troubling who was the theoretical ruler. Once he saw that he had his way at every point, he became accustomed to govern; but if he had been an English or a French general, or one of the military commanders under William I., he would never have dreamt of interfering in politics. It was his fate and his misfortune that he was called upon at a turning-point in German history, when the Kaiser had ceased to play the part assigned to him in the Bismarckian Constitution and a new Constitution had not yet been evolved. It was Ludendorff's ill fortune that he had to fill the gap between two periods of German history.

He was soon to learn the mistake that had been made in appointing a second-rate personality to the office of Chancellor. After the first meeting between Michaelis and the party leaders, von Payer noted in his diary: "We separated under such a cloud of depression that even Bethmann Hollweg's friends failed to derive any satisfaction from the embarrassment of his opponents".

True, Bethmann Hollweg was an intellectual giant compared with his successor. If we could see in the tragedy of a nation arising from civilian incompetence no more than a regrettable error, then this shameful episode in German history would merely merit oblivion. The fact, however, is that the choice of Michaelis, and the readiness to fall down and worship the apparent energy of a bureaucrat, is a specifically German weakness. It is the expert as such who inspires reverence, whether his sphere is the organisation of food supplies or engineering. Michaelis further had a peculiarity which has a twofold influence upon the worshippers of this type: he was not only an expert; to his *expertise* he added moral pathos. The latter is an essential component; the German expert is not satisfied to be master of his trade, but, if he happens to manufacture stoves, he works out a metaphysic of stoves and pretends that God created the world exclusively in order that men might enjoy the gift of stoves. The expert is not content with the usefulness of his activity; he craves to give it a moral halo. Further reflection on this phenomenon will show that it is a first step to that "lust to believe and to obey" in which Nietzsche sees a peculiarity inherent in the German nature.

We know that this deification of obedience is the outcome of the historical development of Germany. Obedience to temporal authority was inevitable, and accordingly, in order to save self-respect, the external "must" was transfigured into an ethical "thou shalt". It was the work of German ethical teachers to make their political misery tolerable to German subjects, and their teachings are the reflection in the realms of metaphysics of a melancholy reality. This is the historical source from which is derived the modern combination of concrete and Kant, of steel constructions and of ethical pathos. Yet it is doubtful whether the expert who sees in the exercise of his profession a kind of high priestly activity is in fact the superior of any clear-minded layman even within his own field. In any case he becomes dangerous as soon as he goes beyond his sphere and seeks to deal with public affairs and world problems with the same confidence which he feels in the exercise of his profession. It is not enough to say that a great and sound people like the Germans fought courageously but at the same time with unparalleled stupidity; and that it "despised the highest of human forces, lacked any internal restraint or plan, worshipped brute force, and thus fell an easy prey to the traps set by a more intelligent if less bellicose enemy".

If the complete lack of intelligence with its endless series of blunders were a natural fact, and somehow organic, it might possess a certain amiability in spite of the harm it does; grandiose in its peculiar stupidity, it might still have been capable of development to something better. But the disastrous negation of intelligence which led to the appointment of a bureaucrat as leader of the nation is of a more complicated nature. It has metaphysical roots; it deliberately denies the supremacy of reason, and rates the possession of expert knowledge of some insignificant subject, coupled with a shopkeeper's business instinct, the two being interpreted as an activity divinely ordained, far above the possession of brains. There can be only one justification for such a type—its success. Fortunately, perhaps, even success was wanting.

During the days of disastrous muddle while Michaelis was conducting the affairs of Germany, Ludendorff wanted to get back to his work. Himself an expert, and succumbing to Michaelis's show of energy, he thought that he had found the man who would support the views of General Headquarters. He now desired to have done with politics and to get back to his real work of conducting the war. It was Michaelis who persuaded Hindenburg and Ludendorff to take a part in politics. "I asked him", Ludendorff writes, "to excuse me; I had the feeling that we should merely be entangled in politics, but the Chancellor remained obstinate." Michaelis's first aim was to keep back the peace resolution prepared by the majority parties. For this purpose, conversations between the party leaders took place at the Home Office; but before they were concluded the text of the resolution had been set up in the press of the *Vorwärts*. Michaelis requested Ludendorff to stop the publication, and Ludendorff in turn asked Herr Südekum, a member of the Reichstag, to assist him. However, on the next morning the whole of Berlin could read the resolution.

The Reichstag dealt with it on the 14th of July 1917 while discussing the war votes; it had already been adopted in committee. It was on this occasion that Michaelis delivered his first and famous speech containing the words: "As I look at it", words which are likely to attach to his name for all time. It is a grotesque element in these melancholy events that the authors of this first public peace gesture never understood the consequences of the words with which the Chancellor destroyed any chances the resolution may have possessed. Indeed, anyone reading this historical speech and noting the applause which it received from the Right and the Left, cannot fail to marvel at the unsuspecting *naïveté* of the German Reichstag. It was enough for Michaelis to proclaim that they would not wage war for a day once peace with honour could be obtained, to call forth loud cries of "hear, hear" from the Centre, the Progressives, and the Social Democrats. Words like "our methods must be those of reasonable negotiations" evoked fresh rounds of honest applause, and the further remark: "peace must be the foundation of a lasting reconciliation between the nations", was rewarded by prolonged cheers. When he came to the words "peace must ensure that the alliance between our enemies does not turn into an economic ring against us", cries of "hear, hear," from the Centre, the Progressives, and the Social Democrats arose, as the shorthand notes inform us.

The honourable members suspected nothing, and it was only the surprised commentaries of the journalists that awakened them to a sense of what had happened. The peace resolution was carried by the votes of the majority party, the new war credits were voted, and the Reichstag went home. On the 25th of July, Michaelis could write triumphantly to the Crown Prince: "The dreaded resolution has been passed by 212 against 126 votes. My interpretation took from it all its danger. When the time comes the resolution will allow us to conclude any peace we like."

Michaelis was mistaken. In its altered form the resolution was useless for a peace of any kind and had become so much waste paper. It would be irrelevant to remark that a peace such as Germany would have consented to at the time would never have been granted by the Allies. What does matter is that the honest straightforwardness of the resolution in its original form was destroyed by Michaelis's interpretation. The part played by this matter by General Headquarters—in other words, by Ludendorff—was wholly different from that of the Chancellor. The Generals objected to the resolution because they did not believe that the enemy wanted peace, and were afraid that it might produce the impression at the front that the war would soon be over. At the same time they avoided a conflict with the Reichstag majority and accepted the situation, confining themselves to a few unimportant amendments. The essence of the resolution remained intact; it was Michaelis who introduced a note of sycophancy and created the suspicion that it might be twisted from its apparent meaning. The excuse brought forward at a later date, to the effect that the crucial words escaped him in the excitement of speaking, is refuted by his letter to the Crown Prince. However modest our estimate of the possible effect of the peace resolution, we must admit that the distortion it underwent distorted the position of Germany in the eyes of the world.

The important fact is that at this moment, when Russia had cast off its autocracy, a vast propaganda was unfolded in Europe, America, Asia, and Africa in favour of democracy. An unequivocal demonstration in favour of peace on the part of the Reichstag might have modified the fierce black and white of the picture in which Germany was commonly depicted as a country of Huns threatening civilisation. Outside Germany it was the Chancellor's commentary that was read; as a result, the resolution was not taken seriously, and thus foreign opinion about the political events in Germany was more correct than that of the Reichstag majority. After all that had happened, the world refused to believe that such a Reichstag could enforce its will. And the world was right. Once the inertness of the Reichstag had allowed things to drift until Michaelis became Chancellor, his initial speech might have brought that body to its senses; for the speech not only rendered the resolution worthless, but also contained an expression which was simply a negation of parliamentary government: "I refuse to give up control". Michaelis added words which made it clear what he meant by this sentence: "I did not consider a Reichstag like the German fit to decide about peace and war on its own initiative during the war". Michaelis's reasons are a matter of complete indifference, although it is easy enough to imagine what they were. After the 19th of July 1917 one course only was open to the Reichstag majority, which was to inform the Kaiser that they refused to collaborate with this Chancellor. This would have cleared the air and would have effected a saving of three months—a useful period in times of war.

The picture of these days is indeed a memorable one. The question whether it was the General or the Politician who was to lead the nation during the war was so formulated as to make it easy for the General to promote himself to the rank of "Dictator". The question of his paramouncy was not even disputed; and surely the General cannot be blamed for knowing what he wanted. The Chancellor knew it equally well, and indeed from the beginning his ambition was to be the General's lackey. The Reichstag was the only body that did not know what it wanted; but even the Reichstag applauded the lackey of the General. When it recognised its error it was too late. It was the misfortune of the German Reichstag that it looked for salvation in a written Constitution, and failed to realise that it is practice and not theory that counts. Instead of grasping the reins of government which the Kaiser dropped, the Reichstag watched the power being abused by a bureaucrat who was pleasing in the eyes of the Generals. The Reichstag itself was satisfied with asking for a change in the Constitution while the guns were thundering at the front. It might have governed, and it allowed itself to be governed.

In his excellent study of historical greatness Jacob Burckhardt discusses among other things the qualities of the "revolutionary general", and points out that among the essential qualities of such a general there is an intellectual power and ease extending to the functions of apprehension as well as of action, of analysis as well as of synthesis. Ludendorff lacked this gift of apprehension, and was unable to penetrate beyond appearances; but even Ludendorff perceived that an assistant wholly devoid of talent could be of no use to him. It is a poor testimonial to Michaelis that Ludendorff gives him when he remarks that the internal evolution of Germany afforded no room for the development of personalities. With all its limitations, the officer caste produced men of character: in the bureaucratic caste they were sadly lacking. "We were wanting in men; our political system brought forth no creative mind." He envies England for possessing Lloyd George, and France for possessing Clemenceau, and regrets that Germany was without a single man with the unity of character, the sure judgment, and the energy of these two. It is another instance of the tragedy of the German officer that he could stand at the transition point between two periods without suspecting that a new Germany was about to be born. Lloyd George and Clemenceau were not the product of an aristocratic and military nor yet of a bureaucratic tradition; the tradition from which they took their origin was an ancient parliamentary one.

It must further be remembered that Clemenceau, whom Ludendorff admired so much, could never have played the part he did unless the solidarity of the military caste, which had prevailed in France up to the Dreyfus affair, had previously been destroyed. The advantages of the two men in whom the Prussian General finds so much to admire do not consist in their militarism or in their rejection of every pacific gesture—what Ludendorff imagines to be subjective heroism was in fact merely the superiority of their objective situation. They had a better view of things than the German military leaders, and, thanks to German blunders, time was working for them: in Michaelis's place they would have displayed civil instead of military courage. At one point, however, Ludendorff was right; if a Lloyd George or Clemenceau had been in control, Ludendorff would never have had the chance of interfering in politics. Although these men appreciated military talent, including that of Ludendorff (as is proved by Churchill's memoirs), the example of Nivelle shows that they soon made an end of any dangerous signs of independence in a general. When General Foch, at the London Conference in the spring of 1918, began to go beyond his military competence, Clemenceau roughly interrupted him with the words: "Please be silent. I am the French representative."

To leave the conclusion of a lengthy war to the judgment of a general was dangerous formerly: it is impossible in modern warfare, since here even the most brilliant strategist is incapable of taking in the full position. However dangerous the position, he will always tend to risk one more and yet one more battle; Caesar and Hannibal, Charles XII. and Napoleon, were alike ruined by excessive ambition. It was no personal peculiarity of Ludendorff to pursue the phantom of a victorious peace—it is a peculiarity of his class. In 1866 an expression like "shameful peace" was unknown, but these are the words in which Moltke and the other generals of William I. might have expressed their opinion about the peace which Bismarck had concluded with Austria. If Germany had been ruled in 1870 by men like Michaelis and Hertling, the Austrians and Italians would have been in Munich before Christmas of that year.

Why did Ludendorff in the summer of 1917 feel confident of ending the war by crushing the enemy? After the preliminary operations of the 7th of June in the Wytschaete salient, the great Flanders battle began on the 31st of July, and with it the attempt of the Allies to seize what they imagined to be the German submarine base on the coast of Flanders. "The battles of the Western front", Ludendorff writes, "became more difficult and costly than any the German army had gone through hitherto. Nevertheless, General Headquarters was not in a position to draw troops from the East in order to reinforce those of the West. There was work enough for every man on the Eastern front, and it was essential to retain adequate forces there. Russia and Roumania had to be beaten if we wished to be in a position by 1918 to force a decision in the West by attacking France on land and England by submarines, if it should be the case that the latter had not yet brought about the desired effect. The military position compelled me to carry a heavy burden—a burden almost heavier than I could bear. But it was necessary, otherwise the dangers threatening in 1918 might have become unsurmountable." We thus see that the submarines were still a decisive factor in the military programme; the authorities had waited and were still waiting for them to produce a result. By now, however, the hopes that they might prove the "last" or the "only means" of leading the war to a victorious conclusion were on the wane. Ludendorff was no longer under the delusion that the resistance of the enemy could be broken at sea, and had recognised the necessity of seeking a decision on land by means of a great battle in France. Previously, however, he wished to inflict a defeat upon Russia in order to liberate troops for the great battle impending in the West. Thus the war was to last through a fourth winter and to continue into a fifth summer.

If he could have seen behind the lines of the enemy he would have learnt that time had become their strongest ally. The Flanders offensive which caused such anxiety to Ludendorff was undertaken against Lloyd George's wish and in spite of the reluctance of the political leaders. Even in England, where the generals were not the masters but the servants of politics, they had for once succeeded in having their way in spite of the better judgment of the statesmen. Lloyd George, assisted by the advice of Churchill, shared the latter's profound suspicions of a strategy which knew no better method than to sacrifice hecatombs of human beings to the new forms of mechanised warfare. This was yet another instance where the statesmen were superior to the military experts. The fact that it was possible to destroy by mechanical means every living being in a certain area was bound, in their opinion, to introduce a fundamental change in strategy. It was clear that the wisdom of the pre-war soldiers could no longer cope with the changes brought about by the war. The politicians put greater trust in economic warfare than in the old-fashioned offensives, and they refuted the arguments of the generals, who insisted that defence was no less costly than offence. They wished to wait until the United States should be mobilised, and their numerical superiority should be more definitely assured. Meanwhile it was intended to attempt operations in Palestine against the Turks, and in Italy (by means of British and French reinforcements) against Austria. Churchill warned Lloyd George not to undertake an offensive, and urged him not to give way to the generals lest his colleagues should rise up and devour him. There was something Napoleonic in this civilian's estimates of military operations: he knew how to rate his opponent.

In the same way in which Napoleon took the peculiarities of the Austrian generals in Italy into his calculations, Churchill allowed for the nature and limitations of Ludendorff's talents. He knew Ludendorff's weaknesses, but he could also appreciate the strength of the German military expert: his strength lay in defence. The dangerous heights on the eastern edge of the Ypres salient between Passchendaele and Klercken had been fortified with the help of every device

LUDENDORFF AND THE POLITICIANS

that German science and inventiveness could conceive; the ground was covered with "pill boxes", bristled with machine guns, and protected by barbed-wire entanglements until it was impregnable even after the heaviest bombardment. The railway connections behind the German front were as good as, if not better than, those on which the British offensive rested. The Dutch railways were continually bringing up gravel for the production of concrete; and the one army commanded by Crown Prince Rupprecht comprised three times as many divisions as were required to hold the position at any given moment.

Churchill was proved right. After six weeks of the heaviest losses the maximum advance of the British amounted to four miles. "Soon the rain descended and the vast crater fields became a sea of choking, fetid mud in which men, animals, and tanks floundered and perished hopelessly."

The British generals failed to learn the lesson of this hopeless offensive, and an attempt to remove the Chief of the Imperial General Staff was rendered futile by the disunion of the British Cabinet. The attacks were resumed in August. At the same time the French offensive began with an attack upon the Siegfried position at St. Quentin, which was meant to distract the enemy's attention while the main attack, aiming to break through the German front, was delivered at Verdun. In spite of the heavy German losses, Churchill does not attach much importance to the result of these offensives. The generals, he says, had been allowed to have their way, and they carried through their melancholy experiment to the end. "They wore down alike the manhood and the guns of the British army almost to destruction. They did it in the face of the plainest warnings, and of arguments which they could not answer."

His anger against the stupid obstinacy of the generals led Churchill to overlook the severe strain placed on the German front. The smallness of the ground gained by such vast sacrifices of life and expenditure of ammunition implied inefficiency on the part of the generals; indirectly, however, these operations did succeed in hampering Ludendorff in carrying out his plans.

An attack by way of Düna, near to Riga, seemed particularly desirable because of the proximity of Petersburg; the decisive blow, however, was intended to be struck in Moldavia. The operation against Riga was carried out and the town was captured, but the forces of the West were so heavily engaged that it was impossible to commence the Moldavian campaign. Instead, it was decided to support Austria against Italy. "It was with reluctance", Ludendorff says, "that I gave up the Moldavian campaign, because I considered it more important than any operations against Italy. The former operation might have brought the war on the Eastern front nearer its end. . . . The attack against Italy, on the other hand, might have eased the position in the West, but it was doubtful whether it would succeed in bringing about an internal crisis in Italy. . . . The Italian campaign was decided upon in order to prevent the collapse of Austria-Hungary." Strategy after the event is inevitably wiser than the General before. Critics blame Ludendorff for not having followed the advice of Colonel Wetzell, the head of his Operations Section, when he counselled him to seek an impressive success in Italy with adequate forces.

Undoubtedly Ludendorff underestimated the openings available in Italy: it was yet another instance where lack of political comprehension and political leadership had to be paid for dearly. The fact that mistakes were also made on the other side affords little consolation. If in England the political leaders had had their way, the divisions which perished in the mud at Passchendaele would have been employed in Italy and in the East. The superfluous German divisions were tied up in Russia, and the resolve to remain in Russia as a victor —politically a grave mistake—prevented any strategic plan on a large scale. Such conquests as those of Poland, Courland, and Lithuania fascinated Ludendorff and led him to adopt a plan which drew a million men from the scene of the decisive battle. For the campaign in Italy, Ludendorff detailed rather less than seven divisions, two of which were drawn from the Western front.

The success of the campaign, which led to the disintegration of Cadorna's army and the capture of 200,000 men and

1800 guns in the space of three days, justified Colonel Wetzell and all those who had hoped for a really great and decisive result from a campaign conducted in Italy with superior forces. Churchill remarks that it is an instructive question for strategists to study what would have happened if Germany had originally resolved to continue the attack with twelve to fourteen further divisions, which could easily have been available now that the Russian front had collapsed.

"But Ludendorff was nursing other plans, larger, more ambitious, and, as it turned out, fatal to his country. Already the vast design of the offensive of 1918 had gripped his mind. Italy was but a 'side show', worth perhaps 'the bones of a Pomeranian Grenadier', but never to obstruct a classical theory and the supreme trial of strength against the strongest foe. Yet the falling away of Italy, a people of 40 millions, a first-class Power, from the cause of the Allies at this time would have been an event even more pregnant with consequences than all the triumphs of March 21, 1918. To overwhelm Italy and to sue for a general peace afforded still the surest hope for the Central Empires."

But such a hope would have been illusory even after a complete defeat of Italy; for while the idea of a peace by negotiation enjoyed the platonic friendship of the Reichstag majority, it had nothing but enemies among the real rulers. The treatment accorded to the peace note of the Pope is a crucial instance. The first step was taken by the Pope on the 26th of June 1917, on which day the Nuncio Pacelli called upon Bethmann Hollweg, who at that time was still Chancellor, and delivered a letter addressed by the Pope to the Kaiser. Plainly, Pacelli was desirous of learning the aims of Germany, especially with regard to Belgium. Bethmann Hollweg's reply contained an assurance that Germany was willing to grant Belgium independence; at the same time, however, he insisted upon the necessity of safeguards to prevent Belgium from falling under the political, military, and financial control of England and France. The conversation between the Nuncio and the Chancellor also turned upon the question of Alsace-Lorraine, the latter remarking that if France showed signs of being ready for peace the question of frontier adjustment would probably cause no difficulties. Three days later the Nuncio had an audience of the Kaiser, and returned under the impression that his steps had met with a ready response. On the 1st of August (*i.e.* after the fall of Bethmann Hollweg) a note was delivered at Berlin addressed to the leaders of the belligerent nations, containing general outlines as well as specific proposals for a lasting peace. It was urged that peace was impossible unless the occupied territories were surrendered; consequently Belgium must be evacuated and safeguards must be given to ensure its independence of other Powers. Similarly occupied French territory and the German colonies were to be surrendered.

On the 30th of August the Nuncio communicated the British reply to the papal note. The reply pointed out that there was no chance of approaching peace until the Central Powers had published their aims in an official form, and that even with regard to Belgium the Central Powers had not declared whether they were ready to restore the complete independence of this country and to make good the injuries which it had suffered. Accordingly the Nuncio asked whether the German Government was willing to make a definite declaration about Belgium, and whether it would state what safeguards for the independence of Belgium it would require. Whatever view is taken of the Vatican *démarche*, it certainly gave Germany an opportunity of making an explicit statement about Belgium, which constituted the chief obstacle on the way to peace. At the same time it afforded an opportunity of setting out unequivocally the view taken by the German Government about the peace resolution of the Reichstag.

Both Michaelis and the Foreign Minister, Dr. von Kühlmann, refused to give such a declaration. Kühlmann misunderstood or pretended to misunderstand the magnitude of the surrender asked for, and imagined that a declaration that Germany was willing to surrender Belgium would be equivalent to the surrender of a valuable pawn, although it was plain that nobody would have dreamed of entering into peace negotiations before the evacuation of Belgium. Michaelis adopted the theoretical considerations laid down in the memorandum drawn up by General Headquarters in this connection. "In our opinion", he wrote to Count Czernin on the 17th of August 1917, "considerable economic and military control would be accepted by Belgium as the result of negotiations, since Belgium will eventually grasp the fact that an *Anschluss* with Germany is the best guarantee of a promising future". This view sounded somewhat different from the assurances given by Bethmann Hollweg to the Nuncio. It is not due to Kühlmann, and even General Headquarters cannot have been so blind as to believe that such an incorporation of Belgium could have been reached as the result of negotiation; the memorandum of the Generals was supposed to apply only in case of a complete victory. The source of Michaelis's wisdom will always remain a secret.

Meanwhile the Pope was awaiting the German reply. Was Germany or was it not willing to surrender Belgium? The Reichstag, too, was becoming impatient; it had not remained ignorant of the steps taken by the Pope, and, as the Chancellor did nothing to consult the Reichstag, the main committee met on the 22nd of August in order to discuss the papal note. Unfortunately it became plain once more that the Reichstag's efforts at independence were short-lived; after an impressive beginning, it soon lost courage, and the main committee allowed itself to be deprived of what little powers it had, on formal pretexts. Michaelis insisted that the whole House could not discuss the reply to the Pope, and succeeded in bringing about the appointment of a committee of seven, which he felt certain of being able to control. The committee, consisting of Count Westarp, Stresemann, Fehrenbach, Erzberger, Wiemer, Ebert, and Scheidemann, held its first meeting on the 28th of August. Herr Haussmann, a member of the Reichstag, noted in his diary: "A vital decision is about to be taken; it may actually lead up to peace negotiations. We have waited too long already: at the present moment the papal note exerts a pressure on us, and it is backed by Austria. Our timid statesmen tend to take shelter behind this pressure in order to escape the extremists. A month ago the surrender of Belgium would have been a generous act on the part of Germany, and if Asquith's question of the 26th of July had been answered in an equally dignified form by Germany, then the two countries would have broken the spell, and the step taken by the Pope, however valuable, would be a secondary and not a prime cause." All this is very well, but Michaelis's statesmanship was anything but timid; timidity and diffidence were displayed only by the Left wing of the Reichstag. True, in the committee meeting of the 28th of August, Scheidemann asked for a definite answer about Belgium; no resolution, however, was passed. The seven did not meet again; their second meeting was also their last.

Soon after, von Kühlmann placed the completed text of the reply to the Pope before the Reichstag. The note merely contained general remarks upon the subject of peace. In order to justify this stylistic exercise Kühlmann advanced his usual argument to the effect that Belgium was a valuable pawn which must not be surrendered prematurely. Further, he gave the assurance that all that was necessary would be said in "confidential negotiations" with the Vatican. He also reminded the Reichstag that the peace resolution of the 19th of July would serve as guiding line for the negotiations. With the exception of Scheidemann, the Reichstag in its entirety took the proffered bait. Scheidemann was not satisfied with the reason advanced by Kühlmann and asked for a clear statement about Belgium; Fehrenbach, Erzberger, and Payer, however, swallowed the bait and allowed themselves to be taken in by the illusion that the platonic declarations contained in the resolution rendered a definite statement upon the Belgian question superfluous. Westarp and Stresemann, of course, were well satisfied that the note should remain silent upon Belgium, and eventually five out of the seven sanctioned the generalities which Michaelis and Kühlmann intended to transmit to the Pope. It would be incorrect to say that pressure exerted by Headquarters, the will of the Generals, or Ludendorff's dictatorship trampled down the superior insight and intellectual courage of the representatives of the people: nobody interfered with them, and they were at liberty to pull to pieces the wretched phrases and the quaint reasonings of Kühlmann. But the attempt was never made; a

LUDENDORFF AND THE POLITICIANS

useless and meaningless reply received the sanction of the Reichstag, and Michaelis at last had a free hand.

On the 11th of September a council was held at Bellevue Castle. Besides the Kaiser and the Crown Prince, five Ministers—the Chancellor, Kühlmann, Helfferich, Count Roedern, Capelle—and Hindenburg, Ludendorff, and General von Falkenhausen were present. The Kaiser admitted that he had formerly shared the views of Falkenhausen, who demanded the annexation of everything up to the North Sea. The position, however, had changed, and even the annexation of Belgium was dangerous. The Kaiser pointed out that Cardinal Hartmann had strongly pressed him not to insist upon annexation, urging that in the annexed regions the clergy would prove unreliable, and the Walloons insubordinate. The wishes of the Generals were impossible: the coast of Flanders could not be held without annexations. On the other hand it would be possible to demand compensations for the surrender of Belgium. Such compensations might consist in the elimination of British influences, in economic safeguards, and in the solution of the Flemish question.

Ludendorff's language was entirely military. Looking at things from the point of view of the next war, he insisted that the possession of the line of the Meuse would not suffice to protect the German industrial regions, and that an Anglo-Franco-Belgian army must be held at an even greater distance. The only way to bring this about would consist in close economic union between Belgium and Germany. Such a union in turn would demand a lengthy occupation and the seizure of Liége. It was an illusion to hope that Belgium could ever be neutral.

It might have been thought that the Kaiser's views about the impossibility of annexing Belgium might have satisfied the Chancellor. But the Chancellor had an underling's instinct for the true distribution of power. Trembling before Ludendorff's theoretical programme, he hastened to assure Hindenburg in a letter written the next day that he would make the demands of General Headquarters, "whose surrender was unthinkable", part of his plan of negotiation. These demands implied the occupation of Liége and of a certain frontier region until such time as "Belgium should have done everything necessary to safeguard economic *Anschluss*". The Kaiser's decision was passed over in silence, and not even a shadow of the Reichstag resolution remained. Hindenburg's reply to the Chancellor put an end to the assumption that there would be any limit to the occupation of Liége. "The necessary military administrative steps can be taken", he wrote, "only if we are masters of the position at Liége, and continue so. Consequently I cannot imagine that we shall be in a position to leave Liége within the time laid down in any treaty." This language was, at any rate, plain enough.

The reply to the Pope was placed before the Kaiser immediately after the council of the 11th of September. We cannot well blame William II. for sanctioning it; since the Council of Seven allowed itself to be taken in, it is hardly possible to wonder at the credulity of the Kaiser. The note of the German Government was handed to the papal Nuncio at Munich on the 19th of September, and it was only now that the unhappy issue of these events became apparent.

Michaelis replied to Pacelli's letter of the 30th of August on the 24th of September 1917. It was this letter which contained the Chancellor's real answer to the papal note. The letter contains the following passage: "At the present stage we are not yet in a position to comply with Your Excellency's desire and to give a definite declaration about the intentions of the Government with reference to Belgium and to the guarantees asked for. The reason does not consist in any objection on principle to such a surrender entertained by the Government. Its extreme importance for the cause of peace is fully appreciated. . . . On the contrary the objection consists in the fact that certain essential preliminary conditions have not yet been fulfilled." This letter, whose true purport was so plainly apparent beneath its tortuous language, settled the fate of the Pope's intervention on behalf of peace. The Kaiser never saw the Chancellor's letter, and Erzberger and Scheidemann insist that they only heard of Michaelis's secret reply to the Pope after the war. Ludendorff, too, was ignorant. In the selected documents published by him in 1920 his comment on the letter is: "Has come to my notice only now".

We see then that things did not happen as they are described in the simple legend about the dictatorship of the Generals. The Kaiser had asked General Headquarters for a memorandum about the future of Belgium. Ludendorff drew it up automatically, in accordance with orders, and Hindenburg supported it. The Kaiser objected on the ground that it was unsuitable for practical politics, while the Chancellor, terrified by the power of the Generals, took it seriously and thought that a peace might be reached, based upon their theoretical maximum demands. In his reply to the Pope he evaded a declaration about Belgium; secretly, however, he let it be understood that Germany would not dream of restoring the full independence of Belgium. Here once again the Reichstag majority proved a failure. In his expert opinion submitted to the parliamentary committee for the investigation of the causes of the German collapse, Dr. Bredt asks why Michaelis kept his reply to the Nuncio a secret, not even placing it before the Committee of Seven. His own comment is: "This is an instance where one feels positively afraid to draw the logical conclusion". Let the reader not be afraid: the Seven were simply duped.

After the war Michaelis blamed Erzberger for the failure of the Pope's *démarche,* thereby contributing not a little to the wave of hate which eventually brought about Erzberger's ruin. The fact was that in the summer of 1917 Erzberger had spoken at the Catholic Congress at Frankfort on the Main and had referred, among other things, to a report addressed by Count Czernin to the Emperor Charles, in which the position of Austria was described as hopeless. Michaelis considered that the publicity thus given to a melancholy truth put an end to whatever desire may have existed among our enemies to conclude peace. "When the enemy heard that Austria threw her sword at our feet in despair, all readiness for peace vanished. The fault was Erzberger's." Anyone who knew Vienna in 1917 would also know that it did not require Erzberger's speech to inform the rest of the world of the real conditions prevalent there; Paris, London, and Rome were better informed than Berlin. Through many invisible channels, from Austrians and from neutrals, and with good as well as with evil intent, a stream of correct information reached the outer world; the idea of ending the war had nothing to do with the Austrian power of resistance but turned entirely upon Belgium. And whatever hopes may have existed of the restoration of an independent Belgium were irrevocably shattered by Michaelis's letter of the 24th of September 1917.

A study of these events affords a profound insight into the methods by which the war was conducted in Germany. The thorough investigation which, after the war, was ordered by the Reichstag in order to discover the causes of the German collapse did not confine itself to throwing light upon the facts, but also discussed the responsibilities attaching to individuals. However useful and even necessary this tracing out of the individual influences during these great events undoubtedly is, it does not go beyond the delimitation of responsibilities and throws no light upon the origins of this tragedy of blindness and incapacity. The position in the summer of 1917 had some resemblance to that of the previous autumn, when the question of submarine warfare, and with it the adhesion of the United States to the Allied cause, was decided. It is not wholly true, and it certainly is not of crucial importance, that two classes, the possessing classes and the poor, stood opposed to each other within the nation at that time, nor can it be said that the more powerful class decided the event. The longing to win importance and general respect, in other words, the idea of a victorious peace, was not the prerogative of any one class. It is true that certain desires of the "Fatherland" party were of an economic nature; but it is equally true that industrials, merchants, and financiers have a general preference for courses of prudence and compromise, for the reason that the pursuit of a decisive victory brings with it the risk of a decisive defeat.

During the wars of liberation of the 19th century the drama of a national uprising was accompanied by an internal revolution; during the Great War the internal conflicts latent in Germany were awakened in a similar manner. At the same time, all the persons in leading positions showed certain common characteristics at moments of misunderstanding and error; at such moments it would seem as though Germans were misled by some disastrous sixth sense, a sense for the unreal. Suddenly all the standards and proportions of the real world vanish, and generals, statesmen, and politicians act as though living in a world of illusion unilluminated by any rational criticism. Whether this world is in fact an illusion or the result of a dangerous excess of education and school training, it is impossible to avoid the suspicion that it could never have found a place in the popular imagination without those stucco figures which had been set up for the admiration of every German in a thousand different shapes since 1871. German history was distorted in order to increase the loyalty of the citizens, and the result was a decline in the sense of realities. Within the narrow sphere of his business the German is careful and exact and disposed to check every figure; when dealing with generalities and with affairs of state and the world, he is carried along by that dangerous sixth sense mentioned above. In this way it is possible for experts to be entirely reliable on their own subject and to be completely ignorant on all other matters.

If political genius consists in appreciating the true position of affairs and the means available for the realisation of plans, without being deluded by appearances or the public opinion of the moment, then the days of Michaelis were more completely devoid of political genius than any in the history of Germany.

Ludendorff's world, like that of his contemporaries, was populated to a certain extent by the stucco figures of his era. There was in him a certain duality. In matters of everyday life he could be as scientifically exact as Napoleon; for example, in investigating the reason why a certain division proved a failure he had no use for moral platitudes and confined himself to empirical methods. Courage is not merely the result of will; it also depends upon the amount of fat and albumen prevalent in the body. Ludendorff gave instructions for a careful analysis of the rations to be made and instructed the chemists to look to the production of substitutes where there was a deficiency of fat and albumen, since fat and albumen were no less important than cannons and machine guns. But at the same time there lived in Ludendorff another man, possessed of the fatal tendency to measure discipline, courage, and all the rest in terms of a rococo heroism.

During the time of Michaelis's ascendancy the battle of Cambrai was in progress. Under the cover of darkness, and taking shelter under the extensive woods of Havrincourt, the British had collected a number of tanks and Cavalry Divisions; a sudden advance was made early in the morning of the 20th of November 1917 without any extensive artillery preparation. By eight in the morning reports reached Ludendorff that the German front had been broken at several points. The Siegfried line, which was the objective of these attacks, was weakly held, for the operations in Flanders had drawn all the available German troops further north. Ludendorff set in motion a number of divisions belonging to the Crown Prince's army and gave instructions for the army of Crown Prince Rupprecht to despatch forces to the north of Cambrai. "The real magnitude of the break-through", Ludendorff writes, "did not become apparent until noon; once I realised it, it caused me intense anxiety." Two or three days had to elapse before any division could reach its destination: a single division required no less than thirty railway trains; the troops must proceed to the various stations, the trains must be in readiness, the troops must entrain, and the trains must be despatched successively—and all this procedure requires time. The British Army Commander did not make full use of his initial advantage; had he done so, it was Ludendorff's opinion that it would have been impossible to limit his success locally. Had the British pursued their success, Ludendorff exclaims, a very different opinion would have been formed about the Italian campaign. The battle of Cambrai continued until the 5th of December 1917, and the German troops eventually succeeded in winning back large parts of the regions they had lost.

The British success at Cambrai had been made possible by the correct use of tanks. For the first time surprise tactics were effectively employed. Colonel Boraston remarks in his

Houghton Mifflin Company, 1932.

32-26381

x-R

no 90 Cur Curden

023.543
L94H4 K

description of the battle that it refuted the ridiculous talk about the superannuated methods of the British in France in 1916 and 1917; in his opinion it is an excellent example of a military innovation and of bold strategy handled by a man of military genius. Churchill is less enthusiastic in his praises. He points out that not only the general ideas but the methods in the execution of this operation had been suggested to the Generals two year ago, but that nobody had paid any attention. The British High Command would not allow the Tank Corps to make a proper use of its machines. The new arm was used merely as an auxiliary; the brunt of the battle continued to be borne by the artillery and the infantry, while the tanks had been condemned to crawl over the battlefield under the barrage of the concentrated German artillery, or else to sink in the mud of Passchendaele. At Cambrai for the first time the tank became an independent weapon and operated over ground not subjected to any previous artillery preparation. Colonel Fuller has recorded that the attack was a great success. As soon as the tanks, followed by the infantry, advanced, the enemy lost his balance completely; those who did not fly in a panic surrendered with little or no resistance. By 4 P.M. one of the most surprising victories in history had been won. The tanks had done all that could be asked of them, and if reserves had been on the spot the success would have been even more striking. The British had broken through the German line on a front of six miles, taking 10,000 prisoners and capturing 200 guns. Their own losses did not exceed 1500.

In his appreciation of the operations Colonel Fuller agrees with Ludendorff in judging that there was a considerable success which was not properly exploited. In their judgment of the tanks, however, these two authorities are not at one. It had been shown at Cambrai that the tank constituted a weapon which, if properly used, no longer made an attack impossible even against the resources of a defence like that devised by the Germans in 1917. Ludendorff took up a remarkably sceptical attitude. "I intentionally spoke of tank panic," he says, "although this expression was objected to by front-line officers." Their resentment is natural; an appeal to discipline, courage, and nervous resilience was not the right method to deal with a machine. The appeal to moral forces failed before the armoured walls, the motors, and the guns of the tanks, and the German soldier, feeling that his own leaders were failing him technically, repaid them by failing in courage. Churchill, the layman, drew very different conclusions from the battle of Cambrai. He asks whether it would have been impossible to have 3000 tanks in readiness by the spring of 1917, as had been originally suggested to the Generals, and he goes on to ask whether it would have been possible to keep this secret and whether such elaborate preparations could have been concealed from the enemy. To all these questions, he says, there can be only one answer. One-tenth of the work wasted by the Generals upon the preparation of their old-fashioned offensives, one-twentieth of the energy displayed in order to induce their respective Governments to agree to them, and the hundredth part of the losses incurred during these offensives would have smoothed the way for the tanks and would have materially shortened the war. But, he goes on to say, let us suppose that the Germans had discovered the secret. What advantage did Ludendorff derive from this terrifying phenomenon, which was brought to his notice not by rumour or secret reports, but by the tanks themselves? Churchill concludes by saying that it is a melancholy satisfaction to be able to say that, if the French and British commanders were short-sighted, the most capable soldier in Germany obviously was blind, the truth being that all high military experts are members of the same school.

An incapacity to appreciate an innovation is the common characteristic of all experts.

CHAPTER VI

LUDENDORFF AND THE POLITICIANS (*continued*)

The Collapse of Russia—Missed Opportunities—Failure of the Reichstag —Lack of a Peace Programme—Ludendorff and Hoffmann— Threat of Resignation—Diminished Rôle of the Kaiser—The Peace of Brest-Litovsk

It made no difference whether the innovation appeared in the shape of the introduction of tanks or in the Russian Revolution. With the collapse of the old Russia in the autumn of 1917 an enormous perspective opened before anyone possessing eyes to see. What all the experts on Russian history had declared to be impossible had now become a present reality: a real Russian Revolution. Plainly there was here a crisis of the first magnitude. The process of evolution advanced with overwhelming rapidity; developments which formerly took centuries were completed and past within the space of months and even weeks. Ludendorff, the military expert, measured these historic events in the terms of the Prussian drill-book. "Officers lost their privileged position, were deprived of their authority, and were not allowed a higher position than private soldiers; indeed they were soon destined to have an even lower position and to lose all their rights. There were a number of short-sighted people who failed to see that the power of every army and the order of every state is based upon authority. . . . I felt unable to follow the events in Russia with an easy mind. The military position became less difficult, but at the same time fresh dangers arose."

It is a melancholy spectacle to watch Ludendorff failing altogether to grasp the real nature of the Russian Revolution. If there was any occasion when it was made plain how alien politics, ethnology, psychology and, above all, the constructive imagination which is the first qualification of a statesman were to him, it was the present. But Ludendorff was a soldier, and as such the servant of his profession and of its tradition; and by the side of the soldiers there were politicians. Once more and for the last time Germany was favoured by fortune. The Russian March Revolution followed upon the attempts of the Allies to mobilise the Russian people under the banner of democracy and under new commanders, to maintain the alliance imposed upon it, and to save the capital invested in Russia. With the collapse of Kerensky's Government and the emergence of the Bolsheviks there was a complete change. The question was whether the German statesmen would know how to utilise this unhoped-for stroke of luck. Would they conclude peace with the new Russia and demonstrate to all the world that Germany was not fighting for conquests, but merely to preserve her position and her possessions? No promises and no demonstration on the part of the Reichstag could have spoken as clear a language as the conclusion of a peace with unhappy and impoverished Russia on the basis of the *status quo* without any attempt at annexations in Poland, Lithuania, and the Baltic provinces. Eighty divisions were stationed in the East, no less than one-third of the German forces. These would be set free if German statesmanship had the will to set them free by renouncing a policy of annexations in the East. Bredt has pointed out the similarity between the Great War and the Seven Years' War. In the one war Germany, like Prussia in the other, was on the defensive, and in each instance the war was being waged to defend the newly acquired status as a great power. Each time a minor enemy power was invaded—Saxony in the Seven Years' War and Belgium in the Great War—and in each instance the chief enemy was beaten at first—at Prague in the one and in France in the other campaign. The decisive blow, however, failed in each instance, and both Kolin and the battle of the Marne ended in defeat. Thereupon the war dragged on slowly; there are striking victories, as at Rossbach, Leuthen, and Zorndorf, and at Gorlice and Vilna, and in Serbia, Roumania, and Italy respectively; but there were also defeats, as at Hochkirch and Kunersdorf, and at Verdun and Lutzk.

The crisis in each war was brought about by Russia; in the one instance, by the death of the Empress Elizabeth, in the other by the Revolution. Frederick the Great used his last victories, those of Torgau and Burkersdorf, in order to lead up to peace; evacuating Saxony, and giving up all his conquests, he felt that the war had not been waged in vain since it had enabled him to maintain the position of Prussia as a great power. The Germany of William II. did not make a corresponding declaration about Belgium; on the contrary it challenged a new opponent. It was regrettable indeed that, after the Russian Revolution, the much-quoted spirit of Frederick II. failed to inspire German policy.

The fact is that the statesmanship and the intuitive vision were lacking which might have seized and exploited this last fortunate moment. Neither Michaelis nor von Kühlmann nor the party leaders of the Reichstag were the men whom the moment demanded. For a time it seemed as though fortune smiled upon Germany; Michaelis's hours were numbered, and the path was free for any man of talent equal to the opportunity. Michaelis was not swept away by any wave of indignation at the incapacity displayed by him; his term of office was brought to a close by a comparatively trivial affair. After the Reichstag had passed the peace resolution the founders of the "Fatherland" Party, Admiral von Tirpitz and Herr Kapp, a high provincial official, the same who was later to acquire a melancholy notoriety as leader of the "Putsch", had agitated for the spreading of patriotic propaganda in the army. Arrangements of this kind were, in fact, made in the armies in the field, and the soldiers who proved their patriotism in battle had to undergo lectures on patriotism during those periods when they were not in danger of death. The officers who received orders to give these lectures could not be blamed; they did their best. The whole system, however, was an insult to every fighting soldier, besides being a ridiculous piece of amateurishness. Herr Dittmann, a member of the Independent Social Democratic Party, made an attack on this system on the 9th of October 1917. To these complaints a vigorous reply was made in the shape of an attack upon the Social Democrats by Admiral von Capelle, the Minister of Marine, who accused the parliamentary party of fomenting mutiny in the navy and paralysing it in order to enforce peace. Michaelis also spoke and complained of subversive tendencies. The parties of the Left wing, with the exception of Herr Naumann, supported Dittmann, and Herr Ebert actually delivered a speech in which he demanded a breach with the Chancellor and a reform of the whole system of government.

The original cause of the dispute was soon forgotten; it was the general dissatisfaction with the Government which had long been smouldering which now broke out openly. An investigation was ordered which ended in showing that Capelle's accusations were groundless, whereupon the Centre and the National Liberal Party joined the parties of the Left. The *bloc*, which was identified with the peace resolution, succeeded in winning over the National Liberals, thanks to the initiative of Dr. Stresemann; and with the adhesion of this important body the non-party committee of this *bloc* acquired political significance. The question was whether the members of the Reichstag would know how to use their power.

It was a definite step forward when, on the 22nd of October, the committee demanded the resignation of the Chancellor. The demand was strictly constitutional, and fully respected the Kaiser's right to select and appoint a Chancellor. Further, however, a memorandum was submitted to the Kaiser's Civil Cabinet, in which the new Chancellor was required to reach an agreement on internal and foreign policy with the Reichstag. "Before arriving at a decision we request His Majesty the Kaiser to instruct the person selected by him for the office of Chancellor, to enter into discussions with the Reichstag." This was an innovation in Germany; it was the first time that the Reichstag expressed to the Crown its wish to play a part in the appointment of the Chancellor. The fact is that it did not require much courage to make this claim. By now the parliamentary majority embraced all the important parties with the exception of the Conservatives, and there was nobody to oppose their wishes. Certainly the Kaiser would not have opposed them, and he was in no position to do so even if he had been averse to such a western innovation. Nor was any opposition to be looked for from General Headquarters. Ludendorff had perceived that no statesmanship lay behind the strong words which Michaelis habitually used, and felt no desire to uphold the Chancellor. When the majority of the Reichstag demanded the removal of the old Chancellor and desired to have a voice in the appointment of the new one, Ludendorff did nothing.

By this time the Reichstag had obtained the reality of power. The final control of affairs rested with it, and everything depended upon its will, its political instinct, and its power of discovering the right man. At first Michaelis and the Secretary of State, Dr. Helfferich, opposed a certain amount of passive resistance to the wishes of the Reichstag, and when first Herr Trimborn and after him Herr Haussmann clamoured for a reply to the Reichstag memorandum, von Valentini, and after him Helfferich, sought to gain time by urging that the Kaiser must not be hurried in forming any decision. These dilatory measures, however, were in vain.

Michaelis could not but see that his position was untenable. Nevertheless, he attempted to hold it; he was ready to resign from his post as Chancellor, but he insisted upon remaining as Prime Minister of Prussia. Such a compromise was unacceptable to the majority parties: the conflict was not one about the policy to be pursued, nor was the issue one of war and peace; it was a constitutional change that was at stake. The most important demand was that which claimed the introduction of equal suffrage in Prussia. The National Liberals, who had just given their valuable support to the parliamentary majority, were not at one with the parties of the Left in the question of the peace resolution; if they opposed Michaelis it was because they saw in him an enemy of the internal reforms. This conflict could be settled only in Prussia, and the question accordingly was whether Michaelis could remain Prime Minister of Prussia. Michaelis himself took a particular pride in what he considered the memorable act of having opposed the wishes of the majority of the nation. "I claim some credit for having successfully, during my term of office, opposed tendencies towards any changes in the constitution leading to the increase of the powers of the

Reichstag." Thus the parliamentary majority was right in declining to entertain a solution which would have left Michaelis as Prime Minister of Prussia while Count Hertling became Chancellor. Hertling's first mission had failed.

What was to be the next step? The Kaiser might have remonstrated against any restriction of his constitutional rights. This, however, he refrained from doing. It would have been natural if he had hesitated before appointing a Roman Catholic and a Bavarian like Hertling to the office of Prime Minister of Prussia. Such hesitation, however, he succeeded in overcoming. On the 26th of October 1917 the Bavarian Minister, Count Lerchenfeld, once again called upon Hertling to come to Berlin. "I would request you to come at the earliest opportunity. His Majesty will repeat to you his offer of the post of Chancellor, and General Headquarters have promised not to interfere in politics any more. The first step will be for you to reach an agreement with the party leaders. You will receive an enthusiastic welcome: in my opinion it is impossible to decline the offer." General Headquarters did indeed abstain from interfering, although Ludendorff was far from feeling enthusiastic at the choice of Count Hertling, a man of advanced years who, in the days of his prime, had not been his friend. "I did not hear of the call extended to Hertling until it had been definitely formulated. It was possible that Hertling had in the meanwhile formed a better opinion about me. I myself did not know him." Later on Ludendorff added: "When I became acquainted with Count Hertling, I soon became convinced that he was not the man to win the war". But Hertling was acceptable to the Reichstag, and in spite of some misgivings the Kaiser sent for him. General Headquarters remained inactive. It is true that Hertling was far from fulfilling the ideal of a Chancellor such as Germany needed at the moment; but though General Headquarters would much have preferred to see Prince Bülow at the head of the Government, they refrained from interfering in the attempts made by the Reichstag to solve the problem by its own lights. The Reichstag was free and unhampered; there was no autocratic regime to hinder it, and it would be impossible to speak of any "dictatorship of the Generals".

Once more the Reichstag proved a complete and hopeless failure. The demands placed before Count Hertling by the non-party committee were simply childish. The bill for the amendment of the Prussian suffrage was placed at the head of the list, all the remaining items of which were wholly unimportant and partly misleading. It contained such details as a demand for the relaxation of the censorship and of the supervision of political and other associations; an old bill on Labour Boards was resuscitated; finally, the cancellation of certain paragraphs in the Board of Trade Regulations was asked for, and the Chancellor was urged to abide by the outlines drawn up on the 13th of September 1917 and submitted to the Pope. It is impossible to witness without amazement the sudden blindness and demoralisation with which the Reichstag was smitten.

At first, as on previous occasions, it seemed that it would allow itself to be guided by healthy political instincts. It took its courage in both hands, saw the essential point, and, with nobody to oppose it, was in a position to advance unhampered. Then, with its object already in sight, the Reichstag suddenly was struck by confusion, and the inner light was gone. The fact that a war was in progress and that the armies of Germany were entering upon a fourth year of unequal struggle with a world of enemies was forgotten. The vast events in Russia, transforming the map of Europe and the whole course of the war, were out of mind. Millions of German soldiers sent up a daily cry demanding the meaning and end of this senseless slaughter. Meanwhile Honourable Members were concerned with the freedom of political associations. To cancel paragraph 153 of the Board of Trade Regulations was a matter of prime importance, and nothing must stand in the way of the resuscitation of the Labour Boards Bill. As for external politics, these were laid down in those general phrases about peace which the German Government had not felt ashamed to hand to the Pope. However great are the mistakes and errors of which the German military commanders may have been guilty, they are negligible compared with the complete failure of the political civilians of Germany. At General Headquarters, day after day work was being done at the highest pressure; hour after hour attention had to be concentrated upon a thousand points of danger. The politicians were bankrupt of ideas and terrified at every intellectual effort; the Generals, at any rate, were complete masters of their trade. "The German armies fought unbroken; their weapons defended the existence of their country, and the whole world watched with amazement the efforts of the German soldiers and of their leaders. And indeed this was all of which Count Hertling was capable. . . . No attempt was made at any constructive foreign policy until there was a sudden awakening from these slumbers and the terrible facts were revealed." Such is the judgment of Dr. Bredt upon this melancholy chapter of German history.

And indeed it was useless to place any hopes in the Reichstag and the weary old gentleman who was at its head. The very members who had sanctioned the appointment of Hertling, hoping to find in him the man they wanted, confided to their diaries the expression of their disappointment. Thus Haussmann noted: "Hertling soon tires in committee; he is uncertain in everything except his manner of speech. There is no real contact with the Reichstag. Kühlmann is much more popular than Hertling, who had expected that the reverse would be the case." Payer, again, wrote as follows: "Hertling, who comes from Southern Germany, does not feel really at home in the Prussian Cabinet, and cannot but be aware that his Prussian colleagues feel him to be somehow alien". These insinuations, however, are of secondary importance. The essential question was whether Count Hertling would succeed in executing the wish of the Reichstag and transform into practice the renunciation of annexations laid down in the peace resolution. Towards the end of November 1917 an Armistice had been declared on the Eastern front, and on the 22nd of December peace negotiations were entered into with the Bolsheviks. A few days previously, on the 18th of December, a Crown Council had been held at General Headquarters at Kreuznach; however, the ostensible purpose, which was to draw up a political programme for the East at the eleventh hour, was not fulfilled. It should, of course, have been the business of German statesmen to lay down in advance some such common plan to be adopted by all the Central Powers. This had not been done. Bulgaria and Turkey wanted to make annexations; Austria only wanted peace, the position having become so bad that her only desire was to get out of the war as soon as possible with a minimum loss of territory. As for Germany, everything depended upon the political leader of the moment and the man who should advise the Kaiser. It is clear, without any appeal to Bismarck, that it is part of the function of the responsible Minister to effect the transition from war to peace and that in these matters General Headquarters must not be allowed a voice. At the Council of Kreuznach the opposite had been the case. It was Ludendorff who explained the views of General Headquarters, who demanded that the territory of Prussia should be extended in the East by the addition of the so-called Polish safety zone, and who wished to see Courland and Lithuania attached to Germany by means of a personal union with the house of Hohenzollern. At Mitau the Diet, the nobility, and certain representatives of the possessing classes had already offered to the Kaiser the title of Duke, so that Ludendorff could feel assured that the future of Courland would be settled in accordance with his desires. Lithuania, on the other hand, was anything but pro-German; nevertheless, Ludendorff insisted on attaching this country too to the German Empire by means of a personal union. The Kaiser sanctioned the idea of a Prusso-Polish safety zone, and Count Hertling did not contradict him; with regard to the plan for the annexation of Poland and Lithuania, he confined himself to obtaining the concurrence of the heads of the federal states. Accordingly it was on these lines that instructions were given by the Crown Council to von Kühlmann, the leader of the German delegation at Brest-Litovsk. Evidently Count Hertling had forgotten that he was bound by the peace resolution of the Reichstag and that he had signed a document renouncing annexations of any kind. Was this no more than a scrap of paper?

The Russians opened the negotiations at Brest-Litovsk with a resounding appeal. Their manifesto addressed "To All" urged the Allied Powers to join in the Conference under the general motto: "No annexations, no indemnities". The invitation was addressed to the Allied Powers on the 25th of December and a reply was requested within ten days, *i.e.* by the 4th of January 1918. The Austro-Hungarian representative, Count Czernin, accepted the suggestion without reservations, and eventually von Kühlmann also made up his mind to agree. He could do this without running any risk of coming into conflict with the views expressed at the Crown Council. The Russian collapse had engendered no comprehensive plan for the East in Germany, and there was no creative mind at work which could have led England to fear any joint designs formed between Germany and Russia. Consequently there was no chance that Lloyd George and Clemenceau, who, in November, had become head of the Government in France, would accept the invitation to Brest-Litovsk; von Kühlmann was safe in following the call "To All".

The only danger resided in the expression "No annexation, no indemnities" contained in the appeal. Here von Kühlmann tried to evade the difficulty by inserting in his reply to Russia the conditional clause: "If the Allies likewise . . ." Thus a peace without annexations was envisaged only if "All" took part in the negotiations. However, this was not clear from Kühlmann's answer, and the Russians could not but believe that they were assured of a reasonable peace. During the period expiring on the 4th of January the Russian delegates intended to report to their Government at St. Petersburg. It is characteristic of the German diplomats that they took no step to explain to the Russians that they had misunderstood the German intentions.

It fell to General Hoffmann, the representative of General Headquarters, to undertake the unpleasant mission of destroying the pleasing illusion under which the Russian delegates were labouring. Hoffmann records that, if the Russian delegates had been suffered to allow the idea to be formed at St. Petersburg, either by the Government or by the populace in general, that the coming peace would guarantee Russia the pre-war frontiers, the only result later could have been an outburst of furious indignation; the Russians would have said that they had been deceived. Accordingly at luncheon, Hoffmann told the real facts to Joffé, who was his neighbour at table. He explained that Germany did not want any annexations, but did ask for the voluntary cession of Poland, Lithuania, and Courland. He pointed out that the representatives of these three countries had already expressed themselves in favour of secession, and urged that it could not be called annexation if this express desire was carried into practice. "Joffé was dumbfounded," Hoffmann says. "Pokrovsky declared with tears of rage that the forcible annexation of eighteen Governments amounted to a dictated peace. The Russians threatened to leave and to break off negotiations; Count Czernin was in despair." The delegates dispersed to Berlin, Vienna, and Petersburg.

In Berlin the Kaiser sent for General Hoffmann and asked for a detailed report, particularly requesting Hoffmann to give his views about Poland. The General's position was difficult. He had always considered the Polish policy of the Central Powers mistaken and did not approve of the safety zone demanded by General Headquarters. "To attach a wide frontier zone with a population of nearly 2,000,000 Poles, as demanded by General Headquarters, seemed to me to be a disadvantage for Germany. The new eastern frontier should have modified and corrected the old frontier in a few insignificant details only. Such military improvements would have added 100,000 Poles to Germany, but not a man more." The Kaiser shared Hoffmann's opinion. On the 2nd of January 1918 there was a Crown Council at Bellevue Castle, attended by Ludendorff and Hoffmann. "I tried in vain", Hoffmann says, "to have a word with General Ludendorff alone, in order to give him details of the report which the Kaiser had instructed me to prepare." During these deliberations Ludendorff urged that the negotiations at Brest-Litovsk should be accelerated as much as possible "with a view to a blow to be struck in the West". He also urged the transfer of troops to the West. The next subject for discussion was the Polish frontier zone, which Kühlmann, in opposition to General Headquarters, now no longer recommended. The Kaiser took a map on which the Polish frontier was drawn as recommended by General Hoffmann. Appealing to the latter, he declared that he had been convinced by the grave objections which had been urged against the views advanced by General Headquarters; accordingly he must withdraw the sanction already given to these plans. Ludendorff objected passionately; he declined to believe that this decision was final and urgently pressed the Kaiser to give another hearing to General Headquarters. The Kaiser closed this "somewhat painful" scene by saying: "I shall await another *exposé* from General Headquarters".

Ludendorff confesses: "I felt that I had annoyed the Kaiser". He now wished to tender his resignation, and actually discussed with General von Lyncker, the head of the Imperial Military Cabinet, his relations with the Kaiser, saying that he felt he no longer possessed the confidence necessary for carrying out his office. At the same time he held himself at the disposal of the Kaiser in any other capacity. von Lyncker appealed to Hindenburg, and the latter requested Ludendorff to give up any idea of resigning. He took it upon himself to settle the difference. The dispute had become known in Berlin, and news of it had been spread under the somewhat obvious form that Ludendorff was dissatisfied with the manner in which the peace negotiations were being carried out. Ludendorff remarks: "This was incorrect. The Kaiser in person had taken up an attitude hostile to me, and such an attitude in the Supreme War Lord was insufferable and incompatible with my feelings of self-respect." Poor Hoffmann, who had brought his opinion before the Kaiser against his will, had to feel the displeasure of Ludendorff. Under the erroneous idea that Hoffmann opposed the plans of General Headquarters with ulterior motives, Ludendorff demanded his dismissal. The Kaiser, however, supported Hoffmann. At a later time he and Ludendorff talked the matter over, and the misunderstanding was cleared up.

Fundamentally it was the old problem which, since 1916, had been crying in vain for solution. It was the old struggle for supremacy between generals and politicians. Can Ludendorff be blamed for having remained simply and solely a soldier, incapable of leaving the sphere of his trade? This hypertrophy of will is a product of the military training, and

LUDENDORFF AND THE POLITICIANS 149

in this respect even the great Moltke did not differ from Ludendorff.

In his *Gedanken und Erinnerungen* Bismarck writes as follows: "The theory applied by the High Command in their dealings with me, and generally taught as part of the theory of war, might be expressed in the following terms: the Foreign Minister is not allowed to have a voice in affairs until the Generals see fit to close the Temple of Janus. But the double face of Janus might serve to remind them that the Government of a belligerent state must look in other directions than the scene of war. It is the function of General Headquarters to annihilate the enemy forces; it is the purpose of war to win peace under conditions conforming to the general policy pursued by the state. To set up and to delimit the aims to be reached by means of war, and to advise the Sovereign upon this matter, is a political function while the war is in progress no less than in times of peace. The methods of warfare must always depend on the question whether the aim was to obtain the result eventually achieved, or more or less than this result; whether it is desired to obtain or not to obtain an increase of territory, and whether it is desired to obtain pledges for the good behaviour of the enemy or not." Moltke and the "demigods", as the higher officers of the General Staff were called, did not make it easy for Bismarck to maintain the supremacy over the military leaders which is the statesman's due. Only the fact that the statesman did eventually prove victorious in this conflict made it possible to bring the war to a successful conclusion. Bismarck was not the man to put the burden of settling the difference upon the King; he fought his own battle, and appealed to the decision of the King merely like a victorious duellist appealing to the umpire to confirm his victory.

During the conflict of 1918 Hertling remained inactive and took shelter behind the Kaiser. After the latter had made his attempt to safeguard the supremacy of the statesmen the Chancellor hurried to the Generals in order to pacify them. Five days after the conflict which had broken out during the Crown Council at Bellevue, Count Hertling wrote to Hindenburg telling him that instructions had been given to Kühlmann not to enter into any discussions with Czernin about the Polish frontier zone. Nobody was dreaming, he said, of placing General Headquarters before a *fait accompli*, and there was no cause for Ludendorff to think any longer of resigning. "I am animated by the firm hope that a satisfactory solution of the difficulty which has arisen will be found, if only in order that the war may be led to a successful conclusion. . . . If by God's gracious help the proposed offensive, under Your Excellency's proved leadership and supported by the heroism . . . decisive success . . . economic interests . . . international position. . . ." A number of meaningless phrases were placed before Ludendorff, who could not be expected to be impressed by this amalgam of trivialities. Ludendorff saw in this document nothing but weakness, submissiveness, and lack of ideas, and framed his answer accordingly.

The reply was not addressed to Hertling, but was contained in a memorandum addressed by General Headquarters to the Kaiser; the verbiage of a Chancellor was passed over in silence, and an aggressive tone was adopted. The memorandum pointed out that the Kaiser himself had given orders for General Headquarters to take a responsible part in the peace negotiations. "Accordingly Your Majesty has made it our right and our duty to take our share in watching to see that a peace is reached worthy of the sacrifices and achievements of the German nation and the German army." After these preliminary remarks expression is given to grave doubts whether the "political leaders" possess the necessary insight and energy to obtain this end. General Headquarters were not of the opinion that the military position was such as to compel Germany to rest content with an unfavourable frontier in the East. Next there comes an important passage which, for the first time, mentions the intended great battle in the West and informs us that this plan had already been sanctioned by the Kaiser. "I cannot but feel that the manner in which negotiations have been carried on at Brest-Litovsk has had an unfavourable influence upon the morale of the army, which is now about to undergo its severest test. We must defeat the western powers in order to assure the position in the world which Germany needs. It is for this purpose that Your Majesty gave orders for the battle in the West to be undertaken, a battle which will constitute the greatest effort made by us during the war, and involving the very heaviest sacrifices. After the events at Brest-Litovsk I cannot but doubt that when peace comes to be negotiated Germany will not obtain the concessions which our position demands and to which our sacrifices entitle us."

Ludendorff, then, did not underestimate the gigantic struggle which was impending; at the same time he also took it for granted that the battle would prove not merely a victory, but a decisive victory. The demand for a dictated peace in the East was justified by the enormous sacrifices and efforts involved in a battle which still remained to be fought. This idea Ludendorff refused to surrender. The thought of the impending battle and of an annihilating defeat of the enemy caused the political leaders to appear more incapable and insignificant even than they were in fact. "The position which has gradually arisen", so General Headquarters wrote to the Kaiser, "is intolerable in view of the impending decision. It is Your Majesty's privilege", so the memorandum concludes, "to decide. We cannot believe that Your Majesty will order men of sincerity, who have proved loyal servants of their country, to lend their authority and names to a policy which their deepest convictions lead them to consider harmful to Crown and Empire. The heavy task which Your Majesty is placing upon the men who will have to prepare and conduct the operations in the West in conformity with Your Majesty's instructions makes it necessary that they should feel certain of enjoying Your Majesty's fullest confidence. Both the armies and their leaders must be upheld by the feeling that the political success will correspond to the military success. I request Your Majesty . . . to come to a definitive decision. Where the safety of the State is at stake, personal considerations for General Ludendorff and myself must carry no weight. Signed: von Hindenburg, F.M."

In the entire history of war this document is unique. Hindenburg and Ludendorff pretend that the issue of a great battle can be definitely calculated. They argue that they are ready to make every effort to plan and execute the battle so as to lead up to a decisive victory, bringing with it the annihilation of the enemy. But they cannot do this unless they have the fullest confidence of the Kaiser, together with some guarantee that a victorious peace shall crown the victory. It is for the Kaiser to decide; if he holds a different opinion he is challenged to choose different generals.

Such a suggestion, however, was impossible. Since the Crown Council of Bellevue something had happened that interfered with Ludendorff's power of logical thought. He had been hurt far more than he would admit, and his desire to rehabilitate his *amour propre* led him to these fantastic aberrations. He was backed up by Hindenburg, who lent the cover of his name to Ludendorff's impossible suggestions. The absurdity of the idea which Ludendorff and Hindenburg suggested to the Kaiser becomes immediately apparent when it is removed from the atmosphere of irritability into more normal surroundings. Let us take the events of 1866 and assume that before the battle of Königgrätz Moltke had felt as much irritated by Bismarck's insistence on a rapid conclusion of the war, and by his avowed intention of giving Austria an equitable peace, as Ludendorff was by Kühlmann's policy. Let us further assume that Moltke went to King William and told him that he could not plan and execute the impending battle unless the King gave him his fullest confidence, meaning by this that he should trust him not only as general but also as politician, so that the King would have to assure him that the peace would be drawn up according to Moltke's and not to Bismarck's wishes, these wishes implying the annexation of Saxony and of the German parts of Northern Bohemia, Moravia, and Silesia. . . . King William would have sent for a doctor.

It was not what was demanded by Ludendorff, but the demand itself which made his request so impossible. Moltke was like Ludendorff in wishing to achieve conquests: both were annexationists, like all generals. King William was dissatisfied with Bismarck's plans for peace, and he too would not have been sorry to see Prussian territory increased. What was impossible in Hindenburg's and Ludendorff's suggestion was the monstrous assertion that they could guarantee a victory if they were assured that a peace should follow to make the victory worth while. Any thoughtful mind would have perceived immediately that there must be an illusion in using a hypothetical victory as an argument in favour of a policy of conquest. The Prussian military writings dealing with this memorandum, and this its sore spot, lay particular emphasis upon the relation between the General and the Kaiser. In so far as these writings are critical at all, they maintain that according to military standards of thought a commanding general cannot claim the measure of individual liberty which, in this instance, Ludendorff and Hindenburg did claim for themselves. It is admitted that it is unsoldierly to pretend that the appeal to conscience is an ethical postulate, and incompatible with Prussian traditions and military discipline as understood in the 20th century; and it is held that freedom of conscience is limited by the authority of the Supreme War Lord. All this, however, is irrelevant. The history of Prussia records instances where breaches of discipline in favour of independent thought were in the best interests of the country. Ludendorff's fault did not consist in a breach of discipline, but in a breach of the laws of logic.

At a later date Ludendorff himself must have realised that it would be difficult to uphold the memorandum addressed to the Kaiser on the 7th of January 1918 in the form in which the Kaiser received it. In the documents of the High Command published by Ludendorff after the war, the last and decisive paragraph of the memorandum is worded differently from the original. Here the penultimate sentence, preceding the question of the Cabinet, runs as follows: "Your Majesty will not ask me to submit plans for operations which are among the most difficult in history unless they are essential for the attainment of definite political and military ends". This is something totally different. The original says that the great battle which is to assure victory must be fought, and that this heavy task "necessitates" the fullest confidence of the Kaiser; the Generals and the army must know that the political success is going to be worthy of the military victory. The second version says: "If the Kaiser does not want a victorious peace, but will remain satisfied with a compromise, it will not be necessary to fight the greatest battle of history".

Such a demand would be quite reasonable on the assumption that Germany had any chance of reaching a reasonable peace with the Allied Powers. This constructive idea, however, is so heavily camouflaged under polemical phrases that it sounds merely like sarcasm. In any case it may be neglected, since the Kaiser never saw any but the first version.

As things now were, it appeared as though the Kaiser had a choice between the policy of a tame peace in the East, as followed by Kühlmann, in which case he would have to do without Ludendorff and Hindenburg and the great battle in the West; and the retention of Hindenburg and Ludendorff, which would mean an end to the tame policy pursued in the East. The Kaiser received the memorandum on the 8th of January, read it, underlined certain passages, and sent it to the Chancellor. On the 9th of January a courier handed the document to von Kühlmann at Brest-Litovsk. Kühlmann in turn replied to the Chancellor, and asked him to treat his answer as strictly personal and confidential.

Kühlmann began by asking what kind of "responsible co-operation" the Kaiser had conceded to the two Generals, pointing out that the only responsibility known to the German constitution was that of the Chancellor. He found himself unable to accept the claim to take a share in supervising the peace negotiations made by the Generals; all that he was prepared to concede was the right to express an opinion. The last word, in any case, must belong to the Kaiser, in whose person military and civil executive power were united. With regard to the Polish zone, the original source of the dispute, Kühlmann adopted Hoffmann's view, and had no difficulty in refuting the assertion that no opportunity was given to General Headquarters to make themselves heard, since precisely Hoffmann was the delegate of General Headquarters. The chief point of the document was the threatened resignation of Hindenburg and Ludendorff, and on this matter, Kühlmann remarked that such a step would be a national disaster. His final conclusion was that the only way to overcome the differences would be talk them over: a very honest suggestion, but one uninspired by psychological insight. The method of debate is a democratic method, and no soldier will ever understand how it can be made a method of government; he orders or he obeys an order, provided that there is somebody to enforce obedience.

Eventually the Chancellor was asked his opinion. In a memorandum addressed to the Kaiser he gave a reply to the Generals and dealt with the question of responsibility. The view he adopted was the obvious one, that the only kind of responsibility known to the constitution was that of the Chancellor. At the same time he admitted that the peculiar position occupied by Hindenburg and Ludendorff made extra-constitutional concessions necessary, although such concessions must not go so far as to allow military counsels to become paramount in political questions. Hence it was inadmissible for Hindenburg and Ludendorff to make a continuation of their indispensable military labours dependent upon the fulfilment of their political demands. If it were admitted that the political views of these two Generals should constitute the last word in politics, this would involve that the entire military and political control should be placed in the hands of these gentlemen. Turning next to the assumption that purely military frontiers could guarantee the security of the Empire, Hertling urged that this type of frontier is the one most calculated to lead to fresh conflicts and wars. In any case annexations were impossible without the sanction of the Reichstag and the Prussian Diet, and accordingly the Chancellor urged the Generals to content themselves with a "true minimum programme". An agreement with the Allied Powers seemed improbable, and consequently "it must be the aim of our policy to maintain friendly relations with Russia and to uphold the alliance with Austria-Hungary". Hertling found himself unable to agree to the demand of the Generals to regard the morale of the army and the nation as a decisive factor; a policy which allowed itself to pursue fleeting successes under the pressure of public opinion would be entering upon a disastrous course. Hertling's memorandum is full of sweet reasonableness; it would have been even more convincing and forcible if the Chancellor had borne in mind obvious parallels drawn from Prussian history. Could it be that Hertling was ignorant of the history of Germany?

The practical result of the effort to clear up and delimit the question of responsibility as between the Chancellor and the Generals eventually resulted in a written agreement by which both parties consented to bind themselves. It was agreed that a division of responsibility was impossible. The Chancellor must have sole responsibility, while the Generals merely had the right to assist at the peace negotiations in a consultative capacity in so far as the negotiations touched upon military matters. The Generals were to advance their demands upon their own initiative, but only in the form of suggestions, advice, and objections. In case of dispute the decision of the Kaiser was to be final. On the 23rd of January Hindenburg and Ludendorff waited upon Hertling, and the submission of the Generals became obvious: the supremacy of the political over the military leaders had at last been recognised. Their submission was also assumed in an autograph letter of the Kaiser's written on the 24th of January to Hindenburg, in which the monarch recognised the "soldierly frankness and outspokenness" of the two Generals, and agreed that they had a right "to give unrestrained expression to their views". At the same time, however, he continued, "the final decision must rest with me. I have passed your memorandum to the Chancellor, and am forwarding to you his observations thereon. I am in agreement with his views, and I expect henceforward that you and General Ludendorff will be able to give up further objections and to devote yourselves wholeheartedly to your proper function of conducting the war...." This language is plain enough; it looked as though the Chancellor's victory was assured. One feels tempted to ask why Hindenburg and Ludendorff composed their memorandum of the 7th January and threatened to resign if it was all to end by their submitting to the political leaders. In his *Memoirs* Ludendorff gives a version of these events in which he tries to make it appear that his main object had been to determine the Chancellor's responsibility in order to prevent that, if any decision were come to of which he, Ludendorff, did not approve, it should be impossible for anyone to pretend that General Headquarters had previously given their sanction. "Count Hertling was obviously at pains to get rid of the imaginary tutelage of General Headquarters. . . . Unfortunately the Government did not make it sufficiently plain in public that it was the Government and not General Ludendorff who was governing."

The fact is that both Kaiser and Chancellor were fully convinced of the necessity of keeping politics free from the influence of the Generals, and that the idea of a dictated peace was distasteful to them, but that at the same time both were too weak to put up a fight for their convictions. If the Reichstag had awakened from its lethargy their position would have been much stronger. The power of the Generals was still based upon the confidence which they enjoyed in the army, in spite of growing dissatisfaction in the nation at large. They knew that the Kaiser was not in a position to dismiss them, and for this reason they attached no undue importance to Hertling's success in the question of responsibility. Soon after they undertook an attack which was particularly unpleasant to the Kaiser, because it was directed against a member of his personal entourage and challenged his undoubted right to choose the head of his personal cabinet. Hindenburg and Ludendorff asked the Kaiser to dismiss the Chief of his Civil Cabinet, Herr von Valentini, a man who had enjoyed his fullest confidence for many years. The Kaiser objected, and asked for an explanation of this peculiar demand.

Hindenburg gave his reply on the 16th of January 1918. Once more he appealed to "public morale" and maintained that this was a matter to which he, as Chief of the General Staff, could not be indifferent. According to him, the sins of omission from which the people were suffering, the incapacity to exploit military successes politically, and the failing respect for the crown were, among other things, a legacy of Bethmann Hollweg's administration, and Valentini was partly responsible for these matters in that he had omitted to inform the Kaiser of the true position of affairs. It was high time, he urged, for the Kaiser to have in his immediate surroundings a man "capable of forming a clear and unbiased view of the position, of setting out the facts bluntly and frankly, and of re-establishing contact with the nation". The Kaiser obeyed, and his faithful servant was dismissed. Indeed it is hardly possible any longer to recognise the Kaiser. The self-confidence of the early years of his reign was gone, together with the assurance which, in earlier days, had given him the courage to dismiss Bismarck and to be his own Minister. He continued to give vent to his feelings in marginal notes on memoranda and documents, but there was a remarkable difference between the proud and temperamental outbursts of the past and the bitter and impotent exclamations of the present.

On the 9th of January 1918 an article appeared in the *Berliner Börsen-Zeitung*, apparently inspired by the Foreign Office, dealing with the alleged persecution of Kühlmann and criticising the patriotic hotheads who were afraid that representatives of the Allied Powers might go to Brest-Litovsk and put a premature end to the war. This well-written article received numerous marginal notes from the Kaiser. Every passage containing an attack by Ludendorff (although the all-powerful Quartermaster-General was never mentioned by name) is marked "Good" or "Excellent". One passage runs as follows: "The public continues to demand a statesman capable of leading the nation. Conditions, however, are not such as to allow any statesman to become great." William II. wrote in the margin: "Correct; either he is unpopular with the Reichstag or with Kreuznach or with both". Kreuznach was the seat of General Headquarters, and meant Ludendorff. The article continues: "Conditions now are such that the Foreign Office no longer is paramount—a preposterous state of affairs". William II. remarked: "Naturally; the Kaiser is ignored by both sides". The melancholy note of these *marginalia* indicates how greatly the power of the Kaiser had declined. It may be mentioned that the Kaiser's view upon the Russian peace was correct. The article further pointed out that events in Russia must not be considered with a view to discovering how far they might offer an opportunity for securing the frontier or making annexations, and went on to say that the correct standpoint was that of "world politics". The Kaiser thoroughly concurred, and twice underlined the passage.

The Kaiser's observation that both sides ignored him showed that he was beginning to appreciate the true distribution of political power, with the Reichstag majority and the Chancellor on one side and General Headquarters on the other: the Kaiser had allowed power to slip from his hands and no longer played a part. A coalition between Kaiser, Chancellor, and Reichstag majority would have been natural, since all three differed from General Headquarters in that they were capable of political thought. During the régime of Bethmann Hollweg the Kaiser and the Chancellor had been at one in their support of a reasonable peace. It is true that at that time the Reichstag allied itself with Ludendorff in order to cause the Chancellor's fall. Again, the Kaiser sanctioned the Reichstag peace resolution. At that time he was the ally of the parliamentary majority, and it was Michaelis, the obedient servant of General Headquarters, and not the Kaiser, who, by his peculiar interpretation, rendered the peace resolution worthless. When the Pope was working for peace the Kaiser was ready to support him; the Reichstag failed, and Michaelis thwarted the movement. In the question of equal suffrage for Prussia the Kaiser immediately gave his sanction and did nothing to oppose the Reichstag majority; it was Michaelis and Hertling who offered passive resistance. At the crucial moment of the peace negotiations at Brest-Litovsk the Kaiser stood by the side of von Kühlmann and supported General Hoffmann. He displayed greater courage than the Chancellor, who was afraid of carrying the peace resolution into practice.

What did the Reichstag do? The time had come to carry its precepts into practice. For the second time the peace resolution was about to be tested, and the Reichstag majority had it in its power to control the course of foreign politics. Erzberger, the father of the resolution, did nothing; the progressive party remained passive, and Scheidemann was the only member who, in the debate of the 6th of January 1918, attempted to make the peace of Brest-Litovsk a really important issue. The debate came to nothing, and the Reichstag resumed its slumbers. The internal paralysis of the majority parties was the result of past errors; they had supported

Bethmann Hollweg's eastern policy, and had agreed to the autonomy of Poland, Lithuania, and Courland. Their idea of autonomy was different from that of Ludendorff and Kühlmann; but Generals, Ministers, and Reichstag majority alike were ready to allow these regions to be torn from Russia.

When the negotiations at Brest-Litovsk were resumed, Trotsky arrived. On the 11th of January he delivered his great speech against so-called national autonomy, which he declared to be a pretext for annexations. Kameniev, Trotsky's brother-in-law, spoke on the following day. Kühlmann had arranged with Count Czernin to call upon General Hoffmann to speak at the next suitable opportunity, and Kameniev's provocative speech was followed by what historians of the novelistic school call Hoffmann's "blow of the fist". On the facts, Hoffmann was in the right; endeavours to reach an equitable peace could only be continued if they were not to be interpreted as signs of weakness. Hoffmann himself tells us: "I spoke sitting and perfectly quietly: I did not raise my voice, and it is a fabrication to say that I struck the table with my fist".

The fact that the Russians used the negotiations at Brest-Litovsk as an opportunity of addressing the world in general caused the Conference to drag on for a considerable period. The greatest blunder of German policy had been to appeal to the right of national self-determination in proclaiming the independence of Poland; it now appeared that this error was beyond remedy and an obstacle in the way of the negotiations. If Poland was not surrendered, Courland and Lithuania equally should not be given up. The differences between the two parties were too great to allow of any agreement. Ludendorff urged the necessity of coming to a decision; he wished to know whether a peace would be reached, since this knowledge was necessary before he could determine the number of divisions to remain in the East. On the 4th of February, Ludendorff, Kühlmann, and Czernin met at Berlin. Peace with the Ukraine, a state at that time merely existing on paper, might be reached any day, and Kühlmann assured Ludendorff that within twenty-four hours of the signature of the peace he would definitively break with Trotsky. When

LUDENDORFF AND THE POLITICIANS 161

the moment had come, however, he hesitated. But an appeal by Trotsky to the German people, in which he asked them to follow the Russian example, gave General Headquarters the opportunity of asking the Kaiser to demand a breach with the Russians.

Thereupon Kühlmann sent Trotsky an ultimatum. Trotsky declined to accept it, but at the same time declared that the war was ended and gave demobilisation orders for the Russian army. These measures did not bring about that clarification of matters in the East for which Ludendorff had asked. On the 13th of February a Crown Council was held at Homburg at which Ludendorff at last had his way; the Armistice with Russia was declared at an end and peace was dictated. "This new display of military power", he says, "seemed undesirable to me; at the same time it was impossible to remain inactive while the enemy was daily growing stronger. Our military position and the food supply was such as to demand a clear situation, and caution was necessary. We were safe in assuming that a peace would be reached; nor were we entering upon a military operation into the void, but upon strictly limited steps." At first Count Hertling and Herr von Payer, the Vice-Chancellor, were reluctant to resume the war against Russia; eventually, however, all adopted Ludendorff's views with the exception of Kühlmann, who declined to modify his opinion that it would be a mistake to take up arms again. At the same time he was willing to co-operate in the event that the Chancellor should side with Ludendorff.

At noon on the 18th of February the German armies resumed operations. Two army groups now advanced, that under General Eichhorn moving towards Dünaburg and Reval, while that under Linsingen advanced towards the Ukraine. The Russians offered no resistance, and the whole of Livland and Esthonia were occupied at the first assault. Thereupon Trotsky expressed his willingness to return to Brest-Litovsk and conclude a definitive peace. Negotiations commenced on the 28th of February, and on the 3rd of March 1918, the Russians signed a treaty dictated by Germany.

Such was the peace of Brest-Litovsk. The parliamentary

M

committee of investigation of 1926 describes the results of this peace as disastrous: never before had the impotence of the Reichstag been demonstrated more clearly before all the world. It was not the case that the Reichstag had no power; what it wanted was the will to act. Innumerable instances are there to prove that the Reichstag could have had power by the side of and even in spite of Ludendorff; what was lacking was the will to power. At the very moment when President Wilson was asking whether the Reichstag was the instrument of the military caste or the mandatory of the German nation, this body remained mute. The dictated peace of Brest-Litovsk was received with jubilation, and only the Social Democrats offered any criticism. Thus Scheidemann said: "While unable to approve the manner in which the peace treaty was brought about, we shall not oppose it in view of the fact that it does put an end to the war in the East".

The objections of the Social Democrats were unheard in the general jubilation. To the world at large, however, Brest-Litovsk demonstrated what Germany meant by peace. Such ideas as that of incorporation and sanction became part of the Allied propaganda; for it was now possible to say that the peace of Brest-Litovsk foreshadowed the fate of Belgium if Germany were victorious. The peace had equally disastrous effects in Germany. The elimination of Russia might have been a great success had it been used as a stage to lead up to an equitable peace; but the peace of Brest-Litovsk was dictated, and, as such, was fruitless. Reichstag and Press kept the nation in the dark, and for a brief period enjoyed the pleasure of having dictated a real victorious peace. "This brief joy was the sole result of Brest-Litovsk."

The apologists of the German eastern policy bring forward the fact that it was the Allies, whose silence, after the appeal before the negotiations, slammed the door in the face of the general peace. This is incorrect; the manner in which the invitations were issued at Brest-Litovsk was such that a general peace congress could not possibly result. It was the Germans and not the Russians from whom the suggestion ought to have come. Churchill admits that if this had been the case a suitable ground for negotiations would have offered itself. Russia was in a state of collapse, Italy at her last gasp, France exhausted, and the British forces bled white; the submarine menace had not yet been overcome, and the United States were 3000 miles away. This was a situation of which skilful leadership in Germany might well have availed itself; the numerous acquisitions made by Germany in Russia, and the hatred and fury with which the latter country was regarded by the Allies, would, in Churchill's opinion, have enabled Germany to make territorial concessions to France and to offer to England the complete rehabilitation of Belgium. It must be admitted that this was Germany's great chance; the correct policy was to restore parts of Alsace-Lorraine to France, to make binding promises of the rehabilitation of Belgium, thus leaving England without a reason for continuing the war, and to make good these concessions at the expense of Russia. But, so Churchill continues, Ludendorff had no desires in this direction. Herein Churchill is mistaken.

It would have required a political genius of Bismarck's calibre to enforce a programme of moderation at this time. A peace on these lines, in the conditions of the moment, would have been equivalent to a great German victory. Nobody, however, dreamed of making the concessions Churchill had in mind; not even the Social Democrats. On the 24th of June 1918 Scheidemann, referring to Alsace-Lorraine, declared in the Reichstag: "It must be made clear to Mr. Wilson that on this point there is no case for reparations. Alsace-Lorraine is German soil and will remain so. What is German will remain German: this much goes without saying." Bethmann Hollweg was the first and the last German statesman who would have been ready to rest satisfied with an adjustment of the Russian frontier, and he had, in fact, discussed this possibility with Pacelli, at a time when the latter was submitting the peace proposals of the Pope. Count Hertling had not only abandoned the peace resolution of the Reichstag; since the Allied Powers had declined the call at Brest-Litovsk he now considered himself free of all obligations towards the Allied Powers and empowered to aim at a peace which, while not identical with Ludendorff's military plans, was even further removed from a programme of moderation. In the spring of 1918 an equitable peace based upon the German minimum programme was, in fact, impossible; not because Ludendorff had no desires in that direction, but because not a man in Germany would at that time have put up with such a programme. Now that the war is over it is easy to see that it was Germany's first and last great chance.

CHAPTER VII

THE BEGINNING OF THE END

Modern Warfare—Plans for 1918—The Battle of March 1918

ENGLAND and France had given up the idea of beating Germany to the ground by means of old-fashioned offensives during the course of 1918. Such moderation was the result of a lengthy struggle between the political and the military leaders. Lloyd George, if only by virtue of his parliamentary position, was far stronger than the German Chancellor; yet even Lloyd George had required time and cautious tactics before he succeeded in removing Sir William Robertson, the Chief of the Imperial Staff, and placing Sir Henry Wilson in his position. Only when this had been done could Lloyd George feel assured that the new classes called up would not continue to be wasted in hopeless offensives in the course of 1918. It was resolved to await the help of America; at the same time the conviction grew that an offensive, if it was to be successful, would require very different methods from those hitherto discovered by the War Office. Modern warfare had taken on a new shape. Formerly war was waged by infantry, artillery and cavalry: during the Great War it soon appeared that these arms would not suffice for the successful continuation of an attack, in view of the enormous growth of the powers of the defence.

There were now four new weapons in existence: machine guns, aircraft, tanks, and gas. It was Churchill, and not the generals, who urged an alteration in the relative strength of these arms. The relative strength before the Great War might be expressed as follows: infantry 40 per cent, artillery 40 per cent, aircraft 10 per cent, cavalry $3\frac{1}{2}$ per cent, machine guns

4 per cent, tanks 2 per cent, and gas one half per cent. Henceforth the proportions were to be as follows: infantry 35 per cent, artillery 30 per cent, aircraft 15 per cent, cavalry one half per cent, machine guns 7 per cent, tanks 8 per cent, gas $4\frac{1}{2}$ per cent. It was hoped that the Allies would receive sufficient reinforcements from the United States to ensure a numerical superiority; but besides this, an attempt was to be made to find a new method which should allow these superior forces to advance without interruption. In this manner victory seemed assured.

In the main this question is a technical problem. It alone, however, would not decide the war: a decision could only be reached if all the Allied forces concentrated their entire resources simultaneously against the enemy. During the whole of the war the battle of the Marne had been the only occasion where this had happened on the Western front. At that time, every division and every available man of the German and the French army was continually and simultaneously engaged and in motion; the front extended from Meaux, eastwards of Paris, to Verdun, curving thence to Nancy and thence again into Alsace, and the entire length of the front was not less than 220 miles. Thereafter the character of the war changed. The armies worked so to speak in shifts, and only 8 to 10 per cent were in action at any one time. Battles succeeded one another, and the strength of the defence was such that the entire available artillery of the total front had to be concentrated upon a limited space in order to prepare and support the attack whenever a further advance was contemplated.

It was no longer possible to deliver three or four battles simultaneously. The war became a siege on a gigantic scale. Great and costly as were the individual operations, no single one of them could ever be decisive, since the entire forces of both sides were never engaged at once; thus the annihilation of the enemy was impossible. During any one of the great battles in the West six, seven, or eight British divisions were engaged in desperate battles; a further half dozen would be in reserve, while the rest of the front remained quiet; 20 divisions would lie in the trenches doing their daily work; another 20 would be drilling behind the front; 20,000 men

THE BEGINNING OF THE END 167

would be attending special courses, 10,000 playing football, and 100,000 would be on leave. On the German side the position was similar. It is plain that a single battle could not exhaust the enemy in these circumstances, and consequently could not lead to his annihilation.

Churchill argued that it was necessary to wait for the Americans. By 1919 the Allied Powers would be superior to those of the Germans, and it would only then be possible to risk a decisive battle where the entire resources of the army—every man and every gun—would enter into action within a definite period.

It would further be necessary to overcome the physical and technical difficulties which hitherto had limited the extent of our operations. If we possessed three times the forces of the enemy something might be achieved; but we would never possess such forces. Again something might be done if we had three or four times as many guns; only it was out of the question that we would ever possess them. It was necessary to have recourse to new methods: new weapons must be found capable of supplementing the old to a much greater degree than hitherto: if British, French and Americans could collect a number of guns, tanks, poison gas shells and bombs, sufficient to render possible two or three simultaneous attacks, equivalent to a general battle on a front of some 200 miles, the course of the war would at last proceed with some semblance of a plan. To bring this about, however, time was necessary; it was decided to spend the year 1918 in resisting the German assault, and not to pass over to the grand attack until 1919.

The British and French accordingly resolved to remain on the defensive for the time being. Meanwhile Ludendorff was busy working out the plans of a battle which was destined to be the greatest in history. Since Autumn he had been at one with Hindenburg in the conviction that the kind of peace which both desired could only be won through a great and decisive victory in the West. It was this victorious peace which had been a bone of contention between Ludendorff and the Kaiser, and between the Chancellor and General Headquarters. The Kaiser had given up his intentions of demanding Liége and parts of the coast of Flanders, and he was also ready for concessions in the East; and the Chancellor, for his part, was equally convinced that the future frontiers of Germany as envisaged by Ludendorff could never be realised; but neither the Kaiser nor the Chancellor was willing to entertain the idea of a peace purchased by a surrender of conquered territory, with special concessions to France.

In any case it was extremely doubtful whether such hopes of peace could be realised at all. The manifestoes of the two paramount Powers, England and the United States, as laid down in the speeches of Wilson and Lloyd George in January 1918, had a different ring from former utterances; and the conditions which Ludendorff's confidant, Colonel von Haeften, brought with him from The Hague showed that the Allied Powers were ready to negotiate. At the same moment, however, Clemenceau made his reply to the Socialist Pacifists: "My foreign and my home policy are identical. My home policy is to fight, and my foreign policy is to fight, and I shall continue to fight. I shall continue if Russia betrays us; I shall continue if Roumania is compelled to surrender; to the last quarter of an hour I shall continue to fight. And the last quarter of an hour will be ours." The real position of affairs, however, was such that the warlike gesture was less important than the pacific intentions behind the warlike armour. But the readiness for peace was passive, and the antagonists could not approach each other unless a gesture was made; and such a gesture could only come from Germany.

The world was still in the dark about Germany's intentions with regard to Belgium. The platonic declarations of the Reichstag could not suffice; indefinite as they were, Michaelis had made their meaning even more indefinite, while Count Hertling was unwilling to recognise them at all. If it was seriously intended to end the war by reaching an agreement and offering a minimum programme, the Kaiser and the Chancellor ought to have asked whether the impending battle in the West was required by the political aims; and if Clausewitz's doctrines had any place in Prussian thought, that principle of his which asserts that there is no such thing as a purely military victory was also valid. The victory which is the aim of war is part and parcel of the political purpose. If an agreement was desired the logical course would have been to adopt a large defensive programme. If it was the conviction of the Reichstag majority that the war was being carried on not for purposes of conquest but simply in order to defend and maintain the German Empire, then the conviction could only be carried into practice if all the forces of German strategy, both theoretical and practical, were devoted to the work of defence. If a number of positions behind the front in France, stretching back from Belgium to the Rhine, had been prepared with the same care and thoroughness which were being spent on the plans for the great battle in the West, then, and then only, the purpose of the war as understood by Germany would have been made plain to the world. If the entire forces of the eastern and western, as well as of the home front, had been concentrated upon this work of defence, the defence would have been of insuperable strength. If appeals had been made to the army and the nation, stating clearly that Germany was fighting only to assure a just peace and the preservation of her frontiers, the will of the nation would have been united and its defensive force assured. But such plans and views were never even discussed. The Kaiser had sanctioned the resolve of General Headquarters to deliver the great battle in the West; in military parlance, he had given orders for the battle; and both the Chancellor and the Reichstag had given their agreement.

In his way Ludendorff was consistent. He desired the attack in the West because his aims could not be achieved otherwise than by the annihilation of the enemy. He also knew that the victory for which he hoped would require the most gigantic efforts, since nothing less was needed than the destruction of the French and the British armies. Only if these forces were annihilated Ludendorff could await the arrival of the Americans without misgivings. The decision, further, had to be reached in spring; by the summer the Americans would have arrived and it would be too late. Ludendorff had a right to appeal to Clausewitz since his military plans agreed with his political intentions; the Kaiser, the Chancellor and the Reichstag, on the other hand, while they approved the idea of a battle of annihilation, had no logical justification for their attitude.

In spite of this inconsistency a compromise might have been found. The Kaiser and the Chancellor had been won over to the idea of a grand attack in the West, but at the same time they wished to make a gesture indicating their readiness to reach peace by negotiation. The way out would have been for Germany to declare her aims at the beginning of the offensive. These would be stated to be a just peace, the renunciation of all her conquests, and the rehabilitation of Belgium. If the success of the operations was as great as Ludendorff hoped, it was all the more essential that Europe should know the intentions of Germany.

Such a plan would have met with the whole-hearted approval of the army and of the majority of the nation; but neither the Kaiser, nor the Chancellor, nor the Reichstag dreamed of entertaining such a plan. Ludendorff had not left the Kaiser and the Chancellor in the dark. During an audience at Homburg on the 13th of February 1918 he described the impending battle as the greatest military task which any army had ever undertaken. "The operations will be successful", he continued, "only if the High Command is completely unhampered. . . . We must not flatter ourselves that we shall have an offensive like those in Galicia and Italy; there will be a vast struggle, beginning at one point, continuing at another, and lasting a long time." The battle in the West was the last great effort made by Germany, and we see that the unanimity between King, General, and Statesman, previously described as an essential condition of a successful issue, was in existence in this case. The unanimity was, indeed, more comprehensive; the Reichstag, too, agreed to the idea of a final battle intended to bring about a decision by annihilating the enemy. Ludendorff could rest assured that the responsible leaders were at one with him. He had a perfectly free hand and was at liberty to devote himself to his military tasks.

The preparations for this greatest battle of history had begun in the autumn of 1917. On the 11th of November of that year Ludendorff had summoned the Chiefs of Staff

THE BEGINNING OF THE END 171

of the armies commanded by Crown Prince Rupprecht and the German Crown Prince, Generals von Kuhl and von der Schulenburg, to Mons; the nominal heads of these armies were not asked to attend, and Hindenburg, too, did not take part in the deliberations. At that time Ludendorff had merely determined to deliver his attack in the West and to direct the brunt of it against the British. The idea of an offensive against Italy was never seriously considered. Such a plan would have had a meaning only as part of a definite political aim; it was incompatible with a campaign aiming at the annihilation of the enemy. Similar reasons may have caused Ludendorff not to accept the suggestion made by Colonel Wetzell, the head of the Operations Section.

Wetzell's suggestion was to attack the French in preference to the British. His plan was based in the main on psychological considerations; the idea was to attack Verdun, this being the most vulnerable spot of the French. Rightly carried out from two sides, and not from the front, as Falkenhayn had done, such an operation might well meet with success. But although the fall of Verdun might be a severe blow, it could not annihilate the French forces. If Verdun was captured, the French hopes to regain Alsace-Lorraine would be diminished, without any excessive apprehensions being raised in England. The operation would be the correct step to take if Germany were aiming at a peace by negotiation; the attack on Verdun would require definite political leadership and a clear-cut peace programme with moderate aims. But such a programme was not in existence, and consequently Ludendorff's plan held the field.

At the discussions of the three military leaders it was resolved to deliver the decisive battle in the West before the American troops should make themselves felt. Ludendorff suggested an attack beginning at St. Quentin and continuing in a north-westerly direction in order to roll up the British front. There were, however, other plans, all of which had to be carefully considered. The Operations Section and the General Staffs settled down to work. Each plan received a code name. Thus von Kuhl's plan, involving an attack in Flanders, on a line from La Bassée to Armentières, was called St. George

I, while his plan for an attack on Ypres was called St. George II. The code word Mars covered the plan for an attack on a front from Arras to Notre Dame de Lorette. Finally there was the Michael plan, covering an attack on either side of St. Quentin. In January 1918 Ludendorff, accompanied by Kuhl and Schulenburg, made a tour of the Western front. On the 21st of January a decision was reached. The St. George plans were dropped, as the ground in the valley of the Lys was too much dependent on the weather; the Mars plan was abandoned as being too difficult. The only plan remaining was that named after the Archangel Michael.

The following was the purpose and general idea of this plan: To form a picture of Ludendorff's aims we must envisage the Western Scene of the War. In its rough outlines the front extended almost vertically from the French coast to St. Quentin, thence horizontally to Verdun, and thence once more almost vertically through Lorraine and Alsace to the Swiss frontier. The direction of the first sector was north and south, that of the second east and west, and that of the third again north and south. For the decisive battle only the first two sectors could be considered. The middle sector of the French front might be considered a kind of horizontal girder supporting a vertical British girder. Should it be possible to cut off a piece of the British girder at the point of junction of the two girders, all the rest would be in the air; and further, if the Germans succeeded in penetrating far enough through the gap to threaten the British retreat, a possibility would arise of shattering the whole front: pressed against the coast the British troops would be forced hastily to evacuate France. Consequently it was determined to attack the point where the French army joined the British. In selecting a point to attack, a hundred circumstances enter into consideration; but, other things being equal, a great advantage is gained by attacking a point of junction. The history of all wars waged by allied forces shows that each constituent army is animated by its own peculiar egoism, and it is the law of its own egoism which each army follows so long as a unified command is lacking. This was the case of the armies of the Holy Alliance; it was the same with the German and Austrian armies in the East, and it was not otherwise as between French and British. Consequently a General attacking at the point of junction between these armies would know that if he should succeed in breaking through, the French would first of all think of themselves, in other words, of the protection of Paris, while the British would think of their base, in other words, of the French ports. The fact that the Allies were thus dependent on a psychological law increased the chances of a successful break through at the point of junction.

Undoubtedly Ludendorff's plans for the decisive battle were of great scope and boldness. They have met with praise but also with criticism. Hans Delbrück, Ludendorff's severest critic, calls them a strategic idea of the highest quality, Napoleonic in the best sense of the term. On the other hand General von Moser considers that, unlike Ludendorff's operations of 1914 and 1915, the plan of 1918 was lacking in comprehensiveness and boldness as well as in clearness and simplicity; above all, it wanted a plain idea capable of inspiring subordinate commanders and men. The opinion of the Prussian general sounds as though it had been written by a litterateur; Delbrück's praises are unstinted merely because he wishes to throw additional emphasis upon the inner contradiction of the plan by admitting the greatness of the idea and of the failure.

Both these opinions are incorrect and both clearly show the traces of their origin: they are the ideas of students of military history, and are derived from examples drawn from the periods of heroic warfare. The melancholy desolation of a modern battle of material has no place in them. Ludendorff himself corrects the view that he lost sight of the miseries of actual warfare and concentrated upon a single bold strategic aim in drawing up his plans. He says quite simply: "In adopting the Michael plan I was guided by the question of time and by tactical considerations. I was swayed in the first instance by the weakness of the enemy. Further, tactics were more important than pure strategy; strategy indeed was impossible without a tactical success. A strategist who neglects tactical success is doomed to failure from the beginning, a fact which is amply proved by the Allied attacks in the first three years of the War." Ludendorff was correct; a modern battle is something wholly different from its historic prototype. Modern battles have changed no less than modern generals.

The idea that generalship requires genius is a survival from the period when the plan of a campaign and the outline of a battle were of paramount importance. The technical instruments of warfare were limited; battles were decided by rifle, gun, or sabre. The fascination of a General was due to his gift of divining the enemy's plans, to his concrete imagination, and to the boldness of his plans. The designs of Charles XII., Napoleon, and Moltke were not without an element of genius; and the combination of philosophy and practical strategy contained in the work of Clausewitz, in which the idea of a Cannae, originally due to the Semite Hannibal, is clothed in a Prussian dress, is a part of the history of human thought no less than of wars. Such was the oil with which a General of the old school was anointed. But genius, intuition, and boldness of design and strategic plan were useless when dealing with the hopeless problem of a modern front. It is true that the Great War can be divided into two radically different parts. The battles of the beginning of the War up to the catastrophe of the Marne, the battles in the East, and some of the battles in Italy, had a kinship with those of an older period and bear a similar character. The war of material in the West is the beginning of a new and terrible period. Two gigantic walls possessing maximum defensive qualities stand opposed to each other, and the whole skill of generalship consists in finding a point where a break-through might be effected. To make a hole in the wall was the essential preliminary to any strategic operation, and the mere fact that every general was compelled to begin by attempting a breakthrough made him the slave of tactics. Strategy necessarily had to take a secondary place.

Ludendorff is the last man deserving the criticism of General Moser, who accuses him of a servile attachment to tactics, completely devoid of boldness of design and grandeur of conception. For better or worse, Ludendorff belonged to the school of Schlieffen, from whom he derived his esteem, possibly his over-esteem, of strategical operations, and his steady concentration upon a single great aim. The battle of Tannenberg will always remain one of the greatest examples of a bold strategy; while Ludendorff's plans for the East—the same plans which Falkenhayn thwarted—are unparalleled in military history. Indeed, Ludendorff, in his impetuosity in the pursuit of great aims and of an abstract military idea, bears a certain melancholy resemblance to Hannibal. Both were equally great as generals; both could win battles; and both lost wars. To criticise Ludendorff's motives in adopting the Michael plan is irrelevant; a modern battle compels the general to look for the enemy's weakest spot. At the same time an important aim was kept in view, for, if the plan should succeed, the whole British front might be shaken. It was not, however, this object which decided Ludendorff in attacking at St. Quentin, for strategically this aim could have been reached more quickly in Flanders. The principal consideration was the weakness of the enemy at this particular spot, and the question of time. The Flanders battle would have to be postponed until late in the spring; and Ludendorff was not in a position to wait. The region of St. Quentin, on the other hand, was practicable in all weathers. At the same time this was the enemy's weakest spot. In order to pass over to a battle of movement Ludendorff would first have to break through; but he would have a chance of remaining in motion and fighting the battle to an end only so long as the enemy was not reinforced by American troops. If the break through was to lead to a success, it would have to be undertaken soon. Considerations of time and *terrain* decided everything. This is the simple history of the origin of the greatest battle of all time.

It would thus appear as though Ludendorff, the strategist of the East, had adapted himself to the conditions of the West. He considered the possibilities of the impending battle with the eye of a tactician. Nevertheless he was mistaken; great as he was as a general, he failed to grasp the cruel truth lurking behind the modern front with its myriad artillery defences. In the passage where he remarks that any strategy which fails to bear in mind tactical successes is doomed to failure, he also remarks that this truth can be proved by examples drawn from the Allied attacks. It may be worth while to examine the Allied attacks from this point of view.

Discussions of the German conduct of the war have hitherto omitted to examine the contemporary events on the side of the enemy. Had it been otherwise, the parallelism between the experience of Germany and of the Allies could not have failed to be discovered. On each side there was a struggle between the military and the political leaders; but, while in Germany the political leaders had to struggle in order to assert themselves, and eventually capitulated before the generals, things moved in England and France in such a way that the politicians not only remained dominant in their own sphere, but eventually also encroached upon that of the generals and determined the methods of warfare. This is the fundamental distinction between Germany and her enemies, latent behind an apparent similarity. Germany went to war without political leaders and without a general; France and England were equally without generals, but they did possess political leaders. In Ludendorff Germany discovered a great soldier, while she failed to discover political leaders; in Lloyd George and Clemenceau England and France obtained leaders of first-rate capacity, but they continued to be without great generals. The politicians were compelled to carry on the war practically without assistance, and eventually, in March 1918, they put the right general in the right place. The long struggles and experiments of the English and French politicians resulted eventually in the firm conviction that victory was impossible with the methods and the tactics of "old-fashioned offensives".

A heavy price had been paid for this discovery. Hundreds of thousands of men had to be killed in senseless attacks before it was perceived that the immense advantages of the defence made an offensive in the old style a wholly meaningless undertaking. The discovery had been made after the first battles of material in 1916, and it had been Churchill who had brought forward the most convincing arguments in favour of this truth. Nevertheless the generals succeeded in upholding the old-fashioned methods which they loved in the

THE BEGINNING OF THE END

face of common sense and the better knowledge of the civilians. General Nivelle was the extreme exponent of the insane method of hurling human beings against masses of metal; and indeed, his method had soon to be discarded when entire divisions refused to allow themselves to be sacrificed to a madman. Even now, however, time had to elapse before common sense triumphed, and Lloyd George had to make use of all his skill before he succeeded in ridding himself of the advocates of useless mass murder. A heavy price had to be paid before it was realised that the old-fashioned offensives must, of necessity, prove a failure. Once this discovery had been made, however, the Allies were invincible.

It may be asked what induced Ludendorff to hope that he would be successful where the Allied armies had failed. Why should he succeed in breaking a modern defensive front any more than they? Every soldier knows that a strategic breakthrough requires more than that the front shall simply be pierced at some given point: the hole in the wall is simply the commencement of operations. If the movement comes to a standstill and the enemy is in a position to bring up sufficient reserves to form a new wall behind the broken one, the attack will have been delivered in vain. An advance will have been made, but the new front which has been reached will generally be less satisfactory than the carefully defended front which has been left behind. The hole in the wall at a given point should be no more than the tactical preliminary for a transition to strategic operations in the shape of mobile warfare. It is at this moment that a fresh army is brought up, whose function it is to get the enemy forces on the run at the point where the break-through was effected; an outflanking movement takes place, the enemy communications are cut, the heavy artillery in its emplacements is put out of action, and a new type of warfare commences. All the preparations for a special kind of mechanised warfare and an enormous expenditure of labour and money on the part of the enemy is thereby rendered useless, whereas the attacking army is elaborately prepared for the new type of warfare.

A considerable number of German military writers have

attempted to excuse Ludendorff's resolve to deliver a grand offensive by arguing that attack is the typically German manner of fighting, that a German can demonstrate his superiority only when attacking, and so on. Such arguments are a piece of military lyricism, a survival from the days of heroic warfare, and have a place in mechanised war no more than cavalry charges, military bands and waving flags. Ludendorff himself never paid any attention to such claptrap; his faith in the possibility of the strategic break-through was based upon practical considerations. According to his calculations, it was in the Spring of 1918 that the German armies of the West would for the first time be numerically superior to the enemy, and he was convinced that he would succeed in equipping and training his attacking army until it would be capable of effecting apparent impossibilities. We shall have occasion to record the amazing work done in this direction; but we shall also perceive the limitations of this great military expert of the old school. Ludendorff's preparations for the greatest battle of all times are obscured by a cloud of tragic misunderstanding. He aimed at a maximum; his object was the annihilation of the enemy, and with this end in view he developed his resources to the uttermost degree without suspecting that all these measures did not suffice to overcome the defence in mechanised warfare. A new type of war had arisen and the greatest expert of the old Prussian General Staff was unable to master a battle of this new style.

At the beginning of January 1918, Germany had altogether 231 divisions in the field. Up to the time of the Peace of Brest-Litovsk 151 divisions were fighting in the West and 80 in the East and on other fronts. Of these 80 divisions, 42 were transferred to the West, so that by the Spring of 1918 the number of German divisions in France amounted to 193. The enemy possessed 168 divisions at the end of February 1918, of which 97 were French, 58 British, 10 Belgian, 1 American and 2 Portuguese. There are no figures available allowing us to compare the respective strength of the armies at any one date; the number of divisions does not indicate the actual fighting strength, because in the different armies the divisions were not of equal strength. According to the Ger-

THE BEGINNING OF THE END

man figures, the German army of the West in the middle of March 1918 was 3,574,906 strong, including 3,438,288 other ranks and 136,618 officers. The number of horses was 710,827. The army of the East at the same date contained 1,004,955 other ranks and 40,095 officers; there were 281,770 horses. According to the figures of Captain Wright who was Secretary to the Allied Council of War, the German forces at the end of February 1918 amounted in the West to 1,232,000 rifles, 24,000 sabres, 8800 field guns, and 5500 heavy guns. The allied forces amounted to 1,480,000 rifles, 74,000 sabres, 8900 field guns and 6800 heavy guns. By the middle of March the figures changed in favour of Germany; but the German forces of the West never gained any appreciable numerical superiority.

On the 21st January 1918, Ludendorff definitely adopted the Michael attack, with its assault on either side of St. Quentin. On the 24th the first orders for the battle were issued by General Headquarters. In order to understand the outline for this great operation, it is necessary to form a picture of the German western front as it was before the offensive. The length of the front from the Northern coast of France to Switzerland was 350 miles, and it was held by 10 armies. These armies were divided into three groups. That commanded by Crown Prince Rupprecht of Bavaria extended from the sea to La Fère, which was approximately the point where the front curved from a north and south to an east and west direction. The next group was that commanded by the German Crown Prince and held the whole east and west section on the western front, reaching from La Fère to the plain of the Woevre. Lastly there followed the front of Lorraine and Alsace, held by the group commanded by Duke Albrecht of Württemberg. When the 42 fresh divisions reached the western front, it was necessary to make new arrangements. The army commanded by General Otto von Below, which had taken part in the Italian offensive, was given a position over against Arras and formed part of the army group of Crown Prince Rupprecht. It was designated Army Headquarters No. 17. The army drawn from the East, formerly commanded by General Woyrsch, was taken over by General von Hutier.

It was designated Army Headquarters No. 18, and took a position in front of St. Quentin between the 2nd and the 7th Army. According to the first plan, three armies, the 17th, 2nd, and 18th, were to carry through the battle; according to the arrangement hitherto in force, these three armies belonged to the Army Group of Crown Prince Rupprecht.

These forces Ludendorff proceeded to rearrange. The 18th Army was placed under the control of the Crown Prince, the western wing of whose forces now extended as far as St. Quentin. The eastern wing of this army group was cut short; here a new army group was formed under the command of General von Gallwitz, holding a front from the forest of the Argonne up to the region of Pont à Mousson. Accordingly two army group commanders would be engaged in the forthcoming battle, namely Crown Prince Rupprecht and the German Crown Prince. General von Kuhl was Chief of Staff of the former group, while the control in the latter belonged to General von der Schulenburg. Both these officers, as we saw above, had assisted Ludendorff in working out the plans for the battle. Ludendorff has given us his reasons for placing the 18th Army under the command of the Crown Prince and for allowing two army group commanders to exercise command during the battle. Remembering the Polish campaign of November 1914, he was anxious, he tells us, to assure himself personally of being able to control the course of the battle to a considerable extent. This would be difficult, he points out, if only one army group commander was allowed to give orders, since in such a case any interference from General Headquarters might look like an intrusion on the part of the Higher Command. Further it would be necessary to draw liberally upon the resources of the Crown Prince's Army Group, and this would be easier if there were two army commanders. "I was not influenced by dynastic considerations", Ludendorff says. "I have always been a loyal servant of the King, but at the same time I am an independent man and not a courtier."

On the 10th of March detailed battle orders were issued. The two armies under the command of Crown Prince Rupprecht (the 17th and the 2nd) were to cut off the British

THE BEGINNING OF THE END 181

forces in the Cambrai salient and to seize the line of Croisilles, Bapaume and the mouth of the Omignon. If the attack proceeded satisfactorily, the right wing, consisting of the 17th Army, was to advance beyond Croisilles. Crown Prince Rupprecht's group was further supposed to advance in the direction of Albert and Arras. The left wing, consisting of the 2nd Army, was to hold the Somme, while the right wing (the 17th Army) was to continue the main pressure of the attack in an attempt to shake the British front further in the North. For this purpose all the divisions standing behind the 4th and the 6th Army, the most northerly portion of Crown Prince Rupprecht's army group, were, if necessary, to be drawn upon without delay. The army group commanded by the Crown Prince, the 18th Army, had the Somme and the Crozat Canal for objectives; if it succeeded in reaching them without undue delay, it was further to seize the crossing of the Somme and of the canal. It was further to hold itself in readiness to extend its right wing as far as Péronne. The left wing was to draw reinforcements, if necessary, from the 1st and the 3rd Armies as well as from its immediate neighbour, the 7th.

For the offensive, Ludendorff had set apart 63 divisions, including 52 so-called attacking divisions which had been specially trained and equipped. 1700 batteries of light and heavy artillery—nearly 7000 guns in all—were to turn their destructive fire upon the enemy. The 17th Army contained 19 divisions, the 2nd Army 20, and the 18th Army 24, and apart from these forces three reserve divisions were at Ludendorff's special disposal. The front on which this gigantic attack was to be delivered was 40 miles in length and stretched from Croisilles to La Fère. The distance from Croisilles to the Coast is roughly 60 miles; La Fère, which lies north-east of Paris, can be reached by car in an hour and a half, the distance being some 75 miles. The southern part of this front was covered by the Somme, which flows past St. Quentin and through Ham and Péronne to Amiens. The Crozat canal, which connects the Somme with the Oise, runs parallel to the most southerly part of the position occupied by the German forces before the attack.

Much the most difficult task was that of the 17th Army. Jointly with the 2nd Army, it was to begin by cutting off the Cambrai salient. The course of the front at this point would cause the attack to proceed in a south-westerly direction. After this first attack the 17th Army had orders to change its front, and to attack in the direction of Albert and Arras, in other words, in a north-westerly direction, an extremely difficult manœuvre, since, after the first part of the operation, the artillery would have to take up fresh positions. This second part of the attack, involving as it did the capture of the powerful defences of Arras, would be deprived of the element of surprise; the enemy would have time to bring up reserves in the meanwhile. The task of the 2nd Army was considerably simpler. The initial assault would be straight ahead, *i.e.* in a westerly direction, while in the later stages, after reaching the line Croisilles—Bapaume—Peronne, its left wing would rest upon the Somme and the advance would be continued upon Albert. The 18th Army was to cover the 2nd Army from attacks coming from the South. Its further functions, once the Somme and the Crozat canal had been reached and the crossings seized, were not laid down by Ludendorff.

A most important element in the success of the operations was that of surprise, and the difficulty consisted in making the necessary preparations without the enemy noticing the elaborate movements and concentration of troops and artillery. Since January the railway system, the stations and the roads behind the front had been steadily extended. Telephone lines had to be constructed for the various headquarters and for the artillery; battle headquarters and observation posts had to be arranged and artillery emplacements to be built and camouflaged. Once the attack proved successful and an advance commenced, material for the reconstruction of railways destroyed by the enemy, such as metal, rails, bridge-building material and water tanks, would be required. All these supplies had to be brought up to the front; the work of bringing them up could only be done at night. A special body of officers was detailed to see that the operations were kept secret. So-called security officers carefully observed the troops and censored their letters; the telephones, and all com-

THE BEGINNING OF THE END

munications with the population generally, were strictly supervised. Specialist officers and air force observers with cameras checked the behaviour of the troops by day and by night, and tested the camouflaging of shelters; steps were taken to ensure that no lights were shown, and the Field Security Police was at work in railway trains, at the stations, and in restaurants and estaminets to see that strict secrecy was observed about the German movements.

Another difficulty consisted in training the artillery in new methods of firing. It had been the practice hitherto to preface every infantry operation by several days of bombardment with the object of registering and destroying all objectives. Surprise was rendered impossible by this method; a preliminary practice bombardment by 7000 pieces of artillery would have made obvious the intention and scope of the attack. On this occasion the preliminary bombardment would be quite short. In order to assure success, a new method was employed by which the preliminary registering upon objectives was rendered superfluous. For this purpose large-scale plans were required in which the batteries and trenches of the enemy, as determined by careful triangulation measurements, were shown in detail. Further, the firing of every shot depended upon the temperature and nature of the explosives and upon the influence of the atmospheric pressure and the weather generally on each individual gun. All these data were worked out behind the front and entered in special tables, with the aid of which the gunners were now in a position to register a hit at the first shot. This method was opposed by Ludendorff's own experts; the old artillery officers declined to give up the old-fashioned methods of days of practice firing, and were convinced only by experiments at the artillery school at Maubeuge.

Apart from these preparations, elaborate plans to deceive the enemy had to be worked out for the whole of the Western front. The idea was to be conveyed to him that the main attack was directed against Verdun, and that subsidiary attacks between Rheims and the Argonne, on the Aisne, and in Alsace-Lorraine were impending. Such plans have a two-edged quality. Unless they are carried through with the necessary care and thoroughness they do more harm than good. They must be prepared with the same care as a genuine attack, and the attacking troops themselves must believe that the impending feint is genuine. The final preparatory stage consists in bringing up the troops, supplies, and ammunitions. In the present instance the artillery received four days' munitions, amounting to some 9,000,000 of projectiles for the three armies. The transport of this mass of ammunition required six nights; it took four nights to get the artillery into position.

Detailed instructions had to be given to ensure these night operations being carried through without a hitch. To concentrate an army of this size within a cramped space time-tables worked out to the minute were required. From the end of February until the 20th of March endless columns advanced to the front every night. First of all, infantry and artillery headquarters and intelligence sections took up their positions, followed by divisional headquarters, artillery staffs, engineers staffs, ammunition columns, motor transport columns, the entire artillery, labour companies, aircraft units, balloons, and searchlight trains. Finally the infantry took up their position, together with remount depôts, bridge-building trains, army service corps, medical corps, and field dressing stations. Road transport officers and military police supervised the movement of the troops.

On the 18th of March Hindenburg and Ludendorff arrived at Avesnes; it was from the latter point that Ludendorff intended to follow the battle. On the 20th, as Ludendorff himself tells us, he had to reach the difficult decision whether to attack next day according to plan or to postpone operations. Any delay would make the position of the troops assembled so close to the enemy more difficult. The tension became unendurable.

A further consideration was that of the gas attack. The effect of the gas shells depended on the strength and direction of the wind, and according to the reports of Lieutenant Schmaus, the meteorologist of the General Staff, the wind strength on the morning of the 20th was by no means advantageous. For a moment it seemed necessary to postpone

THE BEGINNING OF THE END 185

the attack. Later on, although the wind still was not particularly favourable, an attack had become possible. At noon Ludendorff gave his final and irrevocable orders.

At 4.40 A.M. on the 21st of March the battle commenced. All was still dark when the 7000 guns assembled on a 40-mile front opened fire simultaneously. It was the most intense piece of artillery fire ever heard by human ears. Mr. Churchill who spent this night on Nurlu, five miles behind the front, at General Tudor's headquarters as the guest of the 9th Division, has described this artillery preparation.

"I woke up in complete silence a few minutes past four and lay musing. Suddenly after what seemed about half an hour . . . exactly as a pianist runs his hands across the keyboard from treble to bass, there rose in less than one minute the most tremendous cannonade I shall ever hear. . . . Far away, both to the north and to the south, the intense roar . . . rolled upwards to us, while through the chinks of the carefully papered window the flame of the bombardment lit like flickering firelight my tiny cabin. . . . The crash of the German shells bursting over our trench lines eight thousand yards away was so overpowering that the accession to the tumult of nearly two hundred guns firing from much nearer to us could not even be distinguished. From the Divisional Headquarters on the high ground of Nurlu one could see the front line for many miles. It swept around us in a wide curve of red leaping flame stretching to the north far along the front of the Third Army, as well as of the Fifth Army on the south, and quite unending in either direction. There were still two hours to daylight, and the enormous explosions of the shells upon our trenches seemed almost to touch each other, with hardly an interval in space or time. Among the bursting shells there rose at intervals, but almost continually, the much larger flame of exploding magazines. The weight and intensity of the bombardment surpassed anything which anyone had ever known before."

Until shortly before 7 the German artillery fired gas shells; the British batteries soon were shrouded in clouds of gas, and their fire ceased. When this bombardment had lasted for two hours, the British trenches, instead of the artillery positions, became the objective, and high explosives were fired in place of gas shells: the field of battle was one roaring sea of mist belching fire. A little before 9 the creeping barrage commenced, and at the same time the German infantry began the assault. The advance was headed by shock troops, light machine guns, *Minenwerfers*, and field artillery, followed by troops in column together with their baggage, engineers, and detachments of the medical corps. Behind these in turn the divisions advanced by battalions, column after column, extending over a distance of 45 miles, accompanied by artillery, ammunition columns, baggage, medical corps detachments, field bakeries and the heavy baggage. After these came the third line, consisting of reserve divisions, infantry in column of route, artillery, engineers, ammunition columns, staffs and cavalry. By this time the infantry troops of the first line had disappeared in the sea of mist, the shells of the creeping barrage whining above their heads, and the smoke and havoc of battle before them. Advancing through the mist they met with advance posts, machine gun emplacements, and strong posts forming the British first line, the whole constituting a complicated system of trenches and tunnels supporting each other. Once this line was passed, they entered the devastated regions of the old Somme battlefields. By noon the German infantry had in many places advanced beyond the British first line, although the British advance posts which survived the German artillery fire and gas attack, and had not yet lost their sense of direction in the fog, offered an obstinate resistance; over the whole battle-front a series of sanguinary man-to-man engagements commenced. On the evening of the first day almost all the British divisions had been ousted from their original battle positions.

The following was the result of the first day as it presented itself to Ludendorff. The 17th Army, commanded by General von Below and constituting the right wing of Crown Prince Rupprecht's army group was attacking the northern front between Croisilles and Cambrai. Here the enemy position was strongest, and the Germans did not penetrate beyond the second line. The creeping barrage, meanwhile, had advanced considerably further, and the infantry had lost touch with it, so that the infantry in the position held at night was without artillery support. Meanwhile the 2nd Army, commanded by General von der Marwitz and attacking between Cambrai and Bellicourt, had penetrated the British second line. On this sector the co-operation between infantry and artillery was more satisfactory. The 18th Army, commanded by General von Hutier, which formed part of the army group commanded by the German Crown Prince, advanced according to plan, and had not stopped at nightfall. It was opposed by the 5th British Army, commanded by General Gough, which had been compelled to evacuate the first line with the loss of several thousands of prisoners and of almost all its artillery.

On the following day, the 22nd, the 17th Army did not advance appreciably. Although its own second line and part of the third line were brought up, it was unable to break the resistance offered by the British second line. The success or failure of the 17th Army decided that of its neighbour, the 2nd. According to plan, these two armies were to co-operate in cutting off the British salient before Cambrai; consequently if the 17th Army failed, the 2nd Army had to advance alone, which in turn left rather too heavy a burden resting on the 17th Army. Ludendorff sums up the position when he says that the army group commanded by Crown Prince Rupprecht had not gained as much ground during the second day between Croisilles and Péronne as the general plan of the battle demanded. The 18th Army, on the other hand, had been extremely successful. Meeting the enemy where he was weakest, its left wing had reached the Crozat canal, capturing 10,000 prisoners and 150 guns; the reserves of this army were intact, and the army itself was still on the advance. This section of the front had only recently been taken over from the French, and General Hutier's 24 divisions were faced by no more than 4 British divisions.

On the morning of the third day, Ludendorff had to decide whether he would abide by his original plan or would modify his designs in accordance with the results actually reached, which would imply an alteration of his plan of battle. His orders were issued at 10.30 A.M.; the 17th Army was to attack, exerting its main pressure in the direction of Arras and St. Pol, while its left wing was to advance on Miraumont. The 2nd Army was to advance upon Miraumont and Lihons. The 18th Army was to advance with most of its forces beyond Ham, the general objective being the line Chaulnes-Noyon. In the afternoon Generals von Kuhl and von der Schulenburg, the chiefs of Staff of the army groups commanded by Crown Prince Rupprecht and the German Crown Prince, were sent for at Avesnes. At the consultation which followed Ludendorff explained his intentions.

A considerable part of the British army had been beaten; the part remaining intact might be estimated at 50 divisions. It was improbable that the French, who possessed 40 divisions, were in a position to undertake an offensive in order to divert the Germans. Ludendorff's aim was to effect a rapid advance on both sides of the Somme and thus to separate French and British. The 17th Army, together with its northern neighbour, the 6th, followed later by the 4th Army, were to attack the British north of the Somme and "to throw them into the sea". The 17th Army was to advance in the general direction of St. Pol, its left wing penetrating via Doullens in the direction of Abbeville. South of the Somme a general attack was to be made on the line Amiens-Montdidier-Noyon, where the French forces were to be pushed back in a south-westerly direction. This plan implied that the 18th Army, which had originally been advancing in a north-westerly direction, would have to change direction and proceed south-west. Further, the 2nd Army was to advance on both sides of the Somme in the general direction of Amiens, remaining in contact with the 18th Army in the process.

We see then that an altogether new plan was now envisaged. Hitherto the main weight of the attack had been carried by the 17th and the 2nd Armies, whose function it was to press back the British in a northerly direction. This was the main element of the battle, and the 18th Army had merely been intended to cover this operation against French interference. Now, however, this latter operation was changed and extended. The 17th, aided by the 6th and the 4th Army, was to attack the main body of the British in a north-westerly

THE BEGINNING OF THE END

direction, the 2nd Army was to advance straight ahead upon Amiens in a westerly direction, and the 18th Army was to advance south-west upon the French. Thus the lines of attack of the three armies radiated from a common point towards north-west, west, and south-west. Ludendorff himself admits that by now the operation had become "very diffuse". On the 25th of March the 17th and the 2nd Army crossed the line Bapaume—Combles, overcoming heavy resistance in the process, while the 18th Army captured Nesle. By now the 17th Army had been fought to a standstill, and although the 2nd Army was still capable of attacking, advance was slow in the shell-pitted regions of the old Somme battlefields; an advance beyond Albert was impossible. The 18th Army continued to advance, and by the evening of the 26th of March Montdidier had been reached.

The actual course of a battle never evolves according to plan: the forces which eventually enter into conflict cannot be calculated beforehand. On the field of battle thousands of incalculable magnitudes contend with each other, and it is the issue of this conflict which decides whether the advance is to continue or is held up, while the general outline of the battle is formed by the sum total of these individual struggles. This greatest of all battles from the very beginning took a course different from that intended by Ludendorff. From the first day it began to develop on lines of its own, and continued contrary to the General's plan and to its own inner laws. It is this kind of development which constitutes the severest test for a general's foresight. According to theory he ought to abide by his original plan without allowing himself to be intoxicated by minor successes; to surrender the strategic idea in favour of tactical gains is supposed to be a sure way to defeat. The question is whether this applies to the modern battle of material. In selecting the point of attack Ludendorff had placed tactics above strategy in order to effect the necessary initial break-through, and similarly he was compelled even now to pursue a tactical success. The first strategic idea of the great attack became invalid at the moment when it was seen that the 17th Army could not fulfil its task. Possibly the fault lay in Ludendorff's original plans;

it might be argued that the enemy resistance facing the 17th Army was underestimated, or that the capabilities of the German forces upon this front were overestimated. In any case Ludendorff was now at the mercy of this unforeseen issue or miscalculation. The strength of a modern defence being such as it is, the success of an attack depends to a very great extent upon surprise; it is this fact, and the comparative immovability of the heavy artillery, which makes it impossible to make good a mistake by a new disposition of the forces, or to develop a plan by strengthening the front of battle where necessary. On the other hand, the 18th Army had unexpectedly found the weakest spot on the enemy front; it advanced and achieved a great tactical success.

It was upon this success that Ludendorff staked everything. The development of the battle had destroyed his old plan; the new plan, which was more comprehensive than the first, was designed to exploit the advance of the 18th Army. Ludendorff's orders of the evening of the 26th of March extended yet further the scope of operations. The 17th Army was ordered, as before, to advance upon Doullens and St. Pol. The 2nd Army, which had hitherto been advancing upon the British, was now to follow the 18th Army in advancing southwest upon the French, while the left wing of the 18th Army was to advance with Compiègne as its objective, with the ultimate hope of reaching Paris. On the extreme left, on the Oise, the 7th Army was to take part in the attack, with the Aisne as its objective; in the extreme north the 6th Army received orders to advance upon Boulogne. Could these aims be reached with the forces at Ludendorff's disposal?

The number of divisions engaged had grown from 63 to 90 and the length of the front from 45 to 90 miles "The enemy was concentrating his forces, and himself passing over to the offensive", Ludendorff admits, "and our own forces were no longer superior to him. The 17th Army could no longer advance. I continued to do my best to strengthen the left wing of the 2nd Army and to direct it, as well as the 18th Army, upon Amiens". During the further course of the battle Ludendorff restricted his aims; he now confined his aims to seizing the important railway junction of Amiens. Here, too, he was

THE BEGINNING OF THE END

destined to fail; the supply of ammunition was insufficient; the construction of railways and roads took time, and in spite of the large stores captured the troops could no longer be rationed properly. "It was abundantly plain that the resistance of the enemy was stronger than our attack." Ludendorff was forced reluctantly to resolve to give up the attack upon Amiens. Gradually the front grew more quiet, with local engagements breaking out here and there. By the 4th of April 1918 the greatest of battles was over.

CHAPTER VIII

GROWING DIFFICULTIES

Retrospect—Criticism—Growing Difficulties—Battle of April 1918—
The Cost of the Offensive

LUDENDORFF sums up the battle of March 1918 by saying that it was a brilliant feat of arms and would always be acknowledged as such. What British and French had failed to achieve the Germans, in the fourth year of war, had succeeded in doing. These words of praise in fact amount to a condemnation. It is a melancholy satisfaction with which General von Kuhl treats the battle in the great investigation which he devotes to it. He sees in it a tactical success: in the course of a few days the Germans had penetrated to a depth of 40 miles into the enemy position; the material captured was enormous, and the German Army communiqué claimed 90,000 prisoners. "We were within an ace of breaking through definitively.... Field-Marshal Haig was ready to retreat to the sea, and General Pétain was already considering measures for saving the capital. Steps were being taken for the evacuation of Paris, and calculations for the transport overseas of the British Army were being made.... According to the testimony of Field-Marshal Haig and General Mangin a few cavalry divisions were all that the Germans needed in order to widen the gap between French and British, thus severing the two armies."

Captain Wright has stated that the Germans were within measurable distance of achieving a final victory. This again may be true; but even if it is true, it is no more than the epitaph upon a great plan. Of the various critics of the German generals, Churchill is the fairest; and Churchill sums up the result of the battle by saying:

GROWING DIFFICULTIES

"The Germans had reoccupied their old battlefields and the regions they had so cruelly devastated and ruined a year before. Once again they entered into possession of those grisly trophies. No fertile provinces, no wealthy cities, no river or mountain barrier, no new untapped resources, were their reward. Only the crater-fields extending abominably wherever the eye could turn. The old trenches, the vast graveyards, the skeletons, the blasted trees, and the pulverised villages—these, from Arras to Montdidier and from St. Quentin to Villers-Bretonneux, were the Dead Sea fruits of the mightiest military conception and the most terrific onslaught which the annals of war record. The price they paid was heavy. They lost for the first time in the war, or at any rate since Ypres in 1914, two soldiers killed for every one British, and three officers killed for every two British. They made 60,000 prisoners and captured over a thousand guns, together with great stores of ammunition and material. But their advantage in prisoners was more than offset by their greater loss in wounded. Their consumption of material exceeded their captures. If the German loss of men was serious, the loss of time was fatal. The great effort had been made and had not succeeded. The German army was no longer crouched, but sprawled. A great part of its reserves had been exposed and involved. The stress of peril on the other hand wrung from the Allies exertions and sacrifices which, as will be seen, far more than made good their losses."

Furthermore, the Allies gained a moral and strategic advantage which the Central Powers had never succeeded in achieving: the two enemy armies, the French and the British, which the German attack had aimed at severing, were now definitively united.

On the 26th March a meeting took place at Doullens between Lloyd George, Clemenceau, Sir Douglas Haig, Sir H. Wilson, the Chief of the Imperial General Staff, Lord Milner, Pétain, and General Foch. In the face of the common danger conflicts had vanished, and Haig declared that the supreme command of the troops from the Alps to the North Sea must forthwith be placed in the hands of Foch. Only a month before, Clemenceau, at a conference in London, had curtly silenced the General with the words: "*Taisez-vous*, it is I who represent France". Now the turn had come for Foch to speak: "You offer me a difficult task. . . . Nevertheless I accept it." It was in this manner that a unified command was created on the Western front. This was the most important result of the German offensive.

There is in existence an entire library of works partly critical and parly eulogistic of Ludendorff's March offensive. Military experts and amateur strategists vie with each other in seeking the reason for his failure. General von Kuhl, who in his capacity of Chief of Staff of Crown Prince Rupprecht's Army Group himself took part in the battle, approves of Ludendorff's plans. The only qualification he makes is that in spite of the careful preparation of the attack, which "was a brilliant piece of organisation and of tactics . . . ., one essential adjunct for a complete break-through" was missing, namely tanks. We have already seen the reason why the value of this important modern weapon was not properly appreciated. The critical investigations of Ludendorff's battle plan all turn around the one point whether his tactical arrangements corresponded with his strategic plan, in other words, whether the distribution of his forces among the three army groups was the correct one. We saw above that Hans Delbrück claims that the plan suffered from an internal contradiction. In his view the main attack, which contained the essence of the strategic idea, was carried through with inadequate forces, while the subsidiary attack, from which a tactical success might be hoped for, was carried through with an excess of forces. Delbrück considers that the only explanation for this contradiction consists in the assumption that Ludendorff himself was not fully convinced of the success of his plans. In carrying through the subsidiary attack with such generous forces he was effecting a kind of insurance which amounted to putting tactics before strategy. The truth is simpler than Delbrück's assumption. Ludendorff was in a position to form a rough opinion of the power of resistance of the various enemy sectors, but he could not know that the 18th Army would find the weakest point on the enemy front. The criticism to the effect that the attack upon the weakest point was carried through with superabundant forces, the lack of which was felt elsewhere, has become possible only through a knowledge of the distribution of the enemy's forces which was acquired after the event. Before the battle Ludendorff was without this information; the most that can be said is that he formed a mistaken estimate of the enemy's resistance; but it is out of place to assert that the German plan contained an inner contradiction.

The decisive cause of the failure is to be looked for elsewhere. Ludendorff's chief assistant, Colonel Wetzell, the Chief of the Operations Section, admits in his memorandum that in view of the numerical strength and the material resources of the enemy there could be no hope of forcing them upon their knees. "On the other hand, there was a hope", he continued, "that an ambitious offensive operation would impress the enemy with our invincibility, and thus make him ready to accept an equitable peace." These words touch the real problem. The German forces available in the West could not suffice to win a decisive battle as envisaged by Schlieffen; to crush the enemy was an impossibility. Delbrück infers that Ludendorff was at fault; knowing the facts he still aimed at annihilating the enemy, in the hope that he might be in a position to dictate terms of peace. This criticism makes no mention of the other factors, of Chancellor, Reichstag, and Kaiser, which also played a part in determining the fate of Germany. In Delbrück's opinion what matters is the real and not the formal responsibility. Ludendorff, he claims, had assumed a decisive voice in questions of war and politics, and should accordingly be held responsible before the nation and before the tribunal of history.

If our investigation were merely a kind of criminal trial and aimed at fixing the guilt upon some definite person, this verdict, evidently, would settle the matter. Historical investigation, however, has no place for such methods. At bottom, Delbrück's opinion merely expresses the old truth that great generals can win battles, but rarely succeed in carrying a war to a successful conclusion. They are soldiers but not statesmen, a truth which appears clearly in the history of Hannibal, Charles XII., and Napoleon. Any investigation of the German failure is bound to be meaningless unless we first understand why the statesmen capitulated before the Generals, and how it was that a representative of the old Prussian General Staff could carry with him not only the majority of the nation and of public opinion, but also the Chancellor and the Kaiser.

The customary explanation asserts that Ludendorff benefited by the traditional deference paid to the military expert. This explanation is inadequate. Prussian history shows us that during its brightest periods no such deference existed: her wars were carried to a successful end by her statesmen in the face of the opposition of the generals. Again, we saw above that it is impossible to uphold the assumption that Imperial Germany perished of the Bismarckian constitution and the attendant impotence of the Reichstag: the Reichstag had fully as much power as the Parliaments of the western Republics.

It is equally untrue to claim that the Kaiser and the Chancellor were unable to withstand the wishes of the supreme command. Even on occasions when it was possible to exert psychological pressure, as the two popular Generals could do by threatening to tender their resignation, the Kaiser and the Chancellor had the power to restore the natural and constitutional paramouncy of the political leaders if they really so desired: in the crisis of January 1918 Ludendorff and Hindenburg had eventually given way. Even more meaningless is the popular formula which speaks of the stronger will of the Generals and of the supreme influence exerted by General Headquarters upon Kaiser, Chancellor, and statesmen. It is true that Ludendorff possessed considerable willpower, and that Hindenburg had a proper sense of his authority, but they were far from being "daemonic" natures, having an infallible hypnotic influence upon the Kaiser, Chancellor, and Reichstag.

What is called will-power in the everyday sense derives its force from a clear understanding of facts. The intensity of the will depends upon the intensity of the conviction that a given course is the only one possible. It was this conviction that was lacking in the men who were supposed to wield political power; the Kaiser, the Chancellor, and the Reichstag equally were devoid of it. It is superfluous to speak of Michaelis as a politician; he rose by the misfortunes of the Reichstag, and greatness was thrust upon him. Again, Count Hertling and the Kaiser were equally without a clear view of facts. If the barometer fell their faint hearts grew dispirited, and a conviction was dimly felt that it might be better to strive after an equitable peace. If the glass rose they bowed before the power of the Generals and dreamt of conquests. The case was similar with the Reichstag. It succeeded in producing the platonic peace resolution; but the resolution, although happily inspired, was a cold enunciation; it was not put forth as a matter of fanatical conviction; and it is such a conviction alone which can convert a scrap of paper into an act of will, and a passive admission into political action.

The position will become plainer if we examine the enemy camp. The Allied fortunes were endangered by their generals to a far greater extent than those of Germany by the German generals. As in Germany, so with the Allies, public opinion, the Press, and considerable sections of Parliament were ready to extend confidence to the generals. But it was precisely at this point that the campaign against the generals commenced. It was carried through with the greatest consistency and deliberateness in England, where the aim of this campaign was not merely to assure the politicians of the last word in the conduct of war; the latter also encroached upon the military field and there too assured themselves of the ultimate control. It was by civilians that an end was put to the senseless offensives, and the supremacy of the politicians was such as to displace the old-fashioned experts from their ruling position. The Prussian generals and the apologists of German strategy when writing their memoirs never fail to say how much they envy the Allies for having succeeded in keeping the ultimate control in the hands of politicians; implicitly they say that all would have been well with Germany if her statesmen had been equally sure of their aims. Such praise is inapposite; it sees only the victorious issue and overlooks the way by which Lloyd George and Clemenceau came to achieve it. In order that this way should be pursued at all, it was necessary first to grasp the fact that victory could not be achieved by means of the methods of the old military school. Lloyd George and Clemenceau decided to make an end of the old-fashioned offensives. It had become technically possible to destroy by mechanical means all life along a limited front or on a given area, and from this fact it followed that a fundamental change in the methods of warfare was necessary. Churchill and Lloyd George, and not the generals, were the first to grasp this truth, and the plans for 1918 and 1919 were drawn up accordingly.

Ludendorff came to the Western front at the time when these radical changes were taking place. In the East, where strategic ideas were more important than tactical considerations, he had had no experience of a modern battle of material. Yet he succeeded in adapting himself to conditions of which he had no previous acquaintance; and the changes in infantry tactics and the new instructions for the artillery, together with his methods of elastic defence and his masterly retreat into the Siegfried line, are examples of his adaptability. This process, however, could not be carried beyond a certain point; once this had been reached, the traditions of the old Prussian Staff College made it impossible for the military expert to appreciate the value of innovations. It is the misfortune of the expert generally that where the transition point between old and new truths is reached the very mass of the knowledge which has grown out of date becomes a dangerous and heavy burden. At this point the amateur becomes superior to the expert; and Churchill, with his mind unembarrassed by ancient military lore, becomes the master of military pedants. Criticisms such as that of Delbrück and other sticklers for details do not reach the essential point. Even if Ludendorff had distributed his forces for the March offensive, as desired by Delbrück, if the 17th Army had advanced more rapidly, and if a few divisions more had been transferred from the East, so that eventually the attack on Amiens had been successful, would all this have sufficed to bring the Allies to their knees? All these arguments are based on the fundamental error which consists in assuming that in the spring of 1918 a decisive battle was possible at all. This is the old heresy of the Prussian Staff

College; the disastrous tendency to consider a historical as equivalent to an eternal truth; the last fading recollection of the heritage of Moltke and Schlieffen.

It remains to ask whether the great March offensive was intended to force a decision, and whether by this gigantic effort Ludendorff had hoped to break the enemy's will to fight. There can be little doubt that the answer is in the affirmative, and General von Kuhl's confirmation is not needed. The General, who had collaborated in working out the plans for the March offensive, says that there can be no doubt that the attack of March 1918 was intended by General Headquarters to decide the Great War and was not merely meant to wear down the enemy gradually by a series of minor operations. In another passage of his work he expresses himself more clearly: "If the offensive failed, the whole war would be lost". These opinions seem to conflict with the views laid down by General Headquarters in justification of the plans worked out subsequently to the March offensive.

These views can best be inferred from a passage, not indeed from Ludendorff's, but from Hindenburg's memoirs: "If we failed in breaking the enemy's resistance at the first blow, further assaults were to follow on different sectors of the enemy front until we succeeded in reaching our aim. . . . On two occasions England had been saved by France in an extremity, and there was hope that on a third occasion we might succeed in finally crushing this adversary. Our operations continued to be guided by the idea of an attack upon the British northern wing. In my opinion the successful conclusion of such an operation would decide the war in our favour." According to Hindenburg's version General Headquarters had all along been considering the possibility of the failure of the great offensive and had decided in this event to continue the attacks elsewhere. This version is in agreement with the facts, inasmuch as Ludendorff, in his memorandum to the Kaiser, had warned him not to hope that the "greatest battle of history" would be in any way comparable with the offensives in Russia or Italy. With such an eventuality in view, Ludendorff, as late as December 1917, had given orders that, besides the Michael attack, the St. George attack was also to be worked out.

In Army Orders of the 10th of February 1918, Ludendorff gave more detailed orders to Crown Prince Rupprecht's Army Group on this subject. St. George was now to be considered as a second act in the progress of the battle for the event that the Michael operation should, instead of leading to a decisive break-through, come to an end in the face of the reserves which the French and British might be expected to bring up. Naturally the St. George plan had to be adapted to such a contingency, the chief result of which would have been that the artillery engaged in the main operation and a large part of the auxiliary arms and material would have to be employed in the "second act". The preliminary details were worked out in February, but the changes in the disposition of the artillery had to wait until the battle had been fought. The second act depended upon the first in other respects than this. The successes achieved on either flank and the alterations in the plan of battle exerted an influence upon the further designs of General Headquarters. While the battle was in progress it remained uncertain whether the second act would proceed according to plan, and the plans themselves remained in the background until the attack came to a standstill before Amiens. Eventually a minor St. George's plan, known under the name of "Georgette", was adopted. This attack commenced on the 9th of April.

The truth is that in working out the plans for the great attack and in delivering the battle Ludendorff was inspired by the hope that he would succeed in deciding the issue of the war. At the same time, however, he did no more than his duty in taking steps for the event that the battle should not meet with the success for which he hoped. Simple as these events may appear, they become charged with significance once we begin to examine the idea by which the subsequent attacks were inspired. A supreme effort had been made, and had been made in vain; could any hope remain that it would be possible to deliver a decisive battle? The break-down of the offensive before Amiens might surely be held to constitute a verdict upon the strategy of annihilation. Ludendorff himself is silent upon this question, and in his memoirs there is only one sentence which allows us to infer that he had looked forward to diplomatic steps on the part of Hertling. Speaking of the impression made upon the enemy by the offensive, he says: "In spite of all my requests, nothing was done in order to turn the victory to account politically". When Hindenburg for his part assumes that it rested with General Headquarters to seek a decisive success elsewhere he is altogether at fault. "On two occasions England had been saved by France in an extremity, and there was a hope that on a third occasion we might succeed in finally crushing this adversary."

There can be no such thing as a strategy of hopes. If in spite of his experiences of the modern battle of material the military expert still believed that a break-through in the strategic sense was possible, the March offensive was the final test of this belief. The German army and the Prussian General Staff had done their utmost; it was impossible to go beyond what they had achieved, and every subsequent operation would necessarily remain behind this supreme endeavour. General Max Hoffmann, Ludendorff's former assistant in the East, charges his chief with failing to have perceived, after the break-through before Amiens had come to a standstill, that it was useless to hope for a decisive victory upon the Western front. "The first attempt, undertaken with all the means at our disposal, had failed, so that it was certain *a fortiori* that further attacks undertaken with diminishing resources could not hope for success." He concludes by saying: "On the day on which Ludendorff broke off the Amiens offensive, it would have been his duty to draw the attention of the Government to the desirability of opening peace negotiations". Hoffmann is right in his opinion that after Amiens it was impossible to win a battle of annihilation, but he is wrong in his inference. The position was not so simple as to permit the General, after the battle had failed, to say: "I and my army have done all that it is in our power to do; there is no chance of victory; it is now for the diplomats to end the war."

General Hoffmann and the critics who share his view overlook the important fact that the March offensive had altered the entire military position. The battle had had a twofold effect. While it did produce an "extreme crisis", as Hindenburg called it, in England and France, it accelerated events in the United States. From April onwards the military machine of the United States worked at a more rapid pace. At the end of December 1917 the number of Americans in France was 150,000, and 25,000 men were landing each month. General Pershing declared that it was impossible to transport an army sufficient for an attack of any importance before the autumn of 1918. Experts like Bacon assure us that in the autumn of 1917 President Wilson had hoped to reach an acceptable peace merely by threats, and that he was convinced of the necessity of engaging the full powers of the United States only by the appeal for help which reached him in March 1918. From that moment the wheels began to turn; and the huge apparatus of armaments, together with the enormous armies which began to be turned out, put an end to all thoughts of peace. "American policy was carried away by the magnitude of its own warlike efforts and propaganda." In May 1918, 240,000 men landed, 40,000 of them being carried in a single transport of 14 vessels. In June and July, 500,000 men reached Europe. The situation is completely misjudged by those critics of the German High Command who assume that there was any possibility of reaching an acceptable peace after the tactical success—and the strategic failure—of the March offensive. The opportunity was gone; the enemy had overcome the "extreme crisis".

The last moment at which peace by negotiation might have been reached was before the March offensive; and even then, in order that such a peace might be reached, it was necessary that a clear-headed statesman controlled strategy as well as politics. If Germany had adopted a peace programme renouncing all her conquests, and a strategy confining itself to a defensive on the large scale, it is possible that she might have remained successful and perhaps even invincible in the final stages of the war. The German acquisitions on the eastern fronts were being guarded by over a million men, and an appeal for the defence of her own frontiers would have met with a response even by a starving nation. But Ludendorff could not fairly be asked to draw up such a programme. The memorandum addressed to the

Kaiser in which Ludendorff laid down the outlines of "the greatest battle in history" contains the aim and objects of this operation. Here Clausewitz's principle of the inseparability of strategy and politics is definitely accepted.

Delbrück's criticism of the March offensive and the subsequent attacks is summed up in the assertion that such a method of warfare would have been "a rational strategy" if the aim had been to inflict a number of heavy separate blows followed by an offer of peace. If the intention was to bring about a great and decisive operation such as Ludendorff desired, then, in Delbrück's opinion, it was "absurd". In this view he is wrong. One and the same military operation cannot, even in theory, have two different objects. The political aim is the decisive element, and it is this which determines the strategy to be adopted. If the political aim in Germany, before the March offensive, had been to make an offer of peace without annexations, a different plan of campaign would have been required. The kind of campaign required might have been of the type suggested by Colonel Wetzell when he recommended an assault on Verdun followed by a number of independent attacks, which might have demonstrated the German power of resistance, but would not have indicated a will to destroy. Ludendorff differed from the political leaders in knowing what he wanted. Up to the time of the March offensive there was an appalling consistency in all his actions.

His declared aim as laid down before the Kaiser and the Chancellor was the annihilation of the enemy; the battle did not seem to him worth risking if a merely equitable peace was to be the result. Accordingly, if the Kaiser had decided to pursue a campaign as outlined by Delbrück, it is improbable that a change at General Headquarters could have been avoided. In no other way, however, could the necessary harmony between politics and strategy be brought about: the Ludendorff workshop declined to turn out any compromise between defensive and offensive. Once the plans for the March offensive were sanctioned the political leaders had to follow the lead of Ludendorff, and with him of the traditional Prussian policy of annihilation. The logic of events continued its way.

On Tuesday the 9th of April the "second act" of the titanic struggle on the British front commenced. Ludendorff had detailed 28 divisions for this battle, belonging to the two northernmost armies of Crown Prince Rupprecht's Army Group, the 6th Army commanded by General von Quast and the 4th Army commanded by General Sixt von Arnim. von Quast had 17 divisions, von Arnim 5, 1 division remained in reserve attached to the Army Group, and 5 were retained at Ludendorff's disposal under the command of General Headquarters. The 6th Army received orders to break through between Armentières and La Bassée on a front of twelve miles, while simultaneously the 4th Army was to attack with the object of capturing the heights north of the plain of the Lys. These heights run east and west from Mount Kemmel to Cassel, south of the old battlefield of Ypres. The front passed east of this city of ruins, forming the well-known Ypres salient. If the Germans succeeded in capturing the heights south of Ypres—Mount Kemmel and the adjacent heights—the British would have to evacuate the salient.

This sector of the British front was not strongly held. The British forces amounted altogether to 58 divisions, of which 46 had been engaged on the Somme. Many had been completely decimated and had undergone heavy losses in prisoners. In the beginning of April, Field-Marshal Haig was engaged on making good his losses and redistributing his forces. No more than 6 divisions stood between the La Bassée Canal and the Ypres Canal, while the front near Neuve Chapelle, where the main weight of the attack of the 6th German Army was to be delivered, was held by a Portuguese division consisting of 4 brigades. A British division was on either side of the Portuguese. The town of Armentières had been reduced to an impassable mass of ruins, having been subjected to a fierce gas bombardment on the 7th of April, and these ruins covered the northern flank of the German 6th Army. On the morning of the 9th of April, 7 German divisions attacked the 4 Portuguese divisions. In the words of Churchill: "No less than seven German divisions fell upon the four Portuguese brigades, and immediately swept them out of existence as a military force."

The Portuguese debacle also affected the British division. Reserves were called upon, but were unavailing to stop the German advance, and fresh masses of Germans continued to pour through at the point where the break-through had been made. On the evening of the first day the Germans had advanced up to Estaires, a distance of some six miles.

At dawn on the following day, the 10th of April, the 4th Army delivered its attack. Arrangements had been made to postpone the attack by twenty-four hours on the assumption that the British reserves would be exhausted by the end of the first day. This assumption was correct, and the attack met with success. By noon Messines and the Wytschaete Ridge had been seized, and before the day was over the whole British front from Wytschaete to the La Bassée Canal was in the hands of the Germans. But although a break-through had thus been effected, on a front of nearly nineteen miles, the enemy flanks held. In the north a Scots division held the Messines Ridge, while in the south, in the region of Givenchy and Festubert, a Lancashire division resisted all attempts to dislodge it. On a small scale the battle pursued the same course as the March offensive.

Here again it is almost frightening to see how the modern battle of material follows its own peculiar laws. The initial successes of a carefully prepared attack carried through with great numerical superiority induced the General to pursue the tactical success of the break-through, to extend the scope of his attack, and to draw upon all available reserves. The neighbouring armies were called upon to give up a number of divisions, and from the 12th of April onwards, 36 instead of the original 28 divisions were engaged. The German front took on a new shape, and at the point where the first success had been achieved it bulged out as far as the forest of Nieppe, twelve miles in advance of the original position. The new salient eventually had a length of forty miles.

We saw how ten days earlier the most furious assaults eventually came to a standstill before Amiens; similarly, on this occasion too a turning-point was reached where the attack died down, reinforcements failed to reach the front line, and the heavy artillery required a pause in order to take up its new position; the enemy resistance grew more powerful, the attackers had to take up the defensive, while the defenders took up the offensive. "Every yard of ground was bitterly contested, and in the furious man-to-man fights which lasted day and night the German losses were at least twice as great as the British. It was here that the real battle of attrition took place; full of dangers, exhausting, and without recognition of mute heroism."

At the beginning of the German attack, British Headquarters had asked for support from General Foch; but the Commander-in-Chief hesitated, and it was only on the 18th of April that the French took over a part of the threatened front, including the Scherpenberg heights and Mount Kemmel. On the following morning these points underwent a heavy German artillery bombardment. The French infantry retired before the enemy gas shells, carrying with them the British forces on their left wing. "Not a man escaped death or capture." With regard to the mutual recriminations after the disaster, Churchill remarks: "There is no doubt that the relations between the French and British commands during the battle period which began on March 21 were not remarkable for a high appreciation of each other's military qualities. The French staff considered that the British had failed and caused great disaster on the common front, and they openly expressed the opinion that the quality of the British troops at this time was mediocre. The British, on the other hand, felt that the help given under a terrific strain had been both thin and slow, and that the entry of French relieving divisions into the battle was nearly always followed by further retirements."

The capture of Mount Kemmel on the 25th of April was the last success achieved by the Germans during this part of the battle. "The taking of Mount Kemmel", Ludendorff says, "was the climax of the German effort. From now onwards increasing numbers of French divisions faced the 4th German Army. Further attacks promised no success. . . . Realising the strength of the enemy, General Headquarters gave orders for the attack to cease."

Military students among the enemy as well as in Germany

have raised the question whether Ludendorff broke off the engagement prematurely. At a later date the French captured the war diaries and archives of the German 4th Army covering the period from the 9th to the 30th of April 1918, a body of documents including the correspondence between Ludendorff and General von Lossberg, the Chief of Staff of the 4th Army. In this correspondence Ludendorff was found to have written: "In view of the steadiness of the defence, the question must be considered of breaking off the attack". Lossberg saw no possibility of crushing the enemy resistance: "With the troops at our disposal the operations offer no hope of success, and it would be better to break them off." There can be no two opinions as to the correctness of this decision. It is true that the enemy was for the moment in a dangerous position, and that, by the time the German attack was broken, they themselves were at the end of their resources. These facts, however, do not invalidate Ludendorff's decision. Churchill's summary is as follows: "From the general, not less than from the British point of view, April 12 is probably, after the Marne, the climax of the war. It looked as if the Germans had resolved to stake their fate and their regathered superiority on battering the life out of the British Army. During twenty days they had hurled nearly ninety divisions in three great battles upon an Army which counted no more than fifty-eight, and of these nearly half were fastened to fronts not under attack. With a superiority of numbers in the areas of assault of three and often four to one, with their brilliantly trained shock troops, with their extraordinary skill and enterprise in manœuvring with machine guns and trench mortars, with their new infiltration scheme, with their corroding mustard gas, with their terrific artillery and great science of war, they might well succeed."

How was it that these successes could not be turned into a decisive victory? The tactical success had been won; the break-through had been achieved; why was it impossible to reach the strategic aim? The reason is that the aim was impossible of realisation. Every attack, however carefully prepared, and however great the numerical superiority with which it was carried through, ended necessarily in a battle of attrition. The modern battle of material does not admit of the decisive battle postulated by Schlieffen.

Again and again we meet with the limitations of the old-fashioned Prussian military theory. This theory was magnificent in its one-sidedness; but it had no place under modern conditions. By this theory the principle of attack was glorified, and the belief in its unique validity was such as to blind men to its merely relative nature. The dialectical methods of Clausewitz were forgotten, and it is no accident that Schlieffen and his zealous disciples, including Ludendorff, have always opposed the fundamental principle in which the philosopher of war refers to defence as the stronger form of warfare. The classical formula states that defence is the more effective form with a negative purpose, and that attack is the less effective form with a positive purpose. The effect of the industrialisation of war was to change the relation between these two forms of warfare in favour of the more effective form: the chances of attack were reduced to nil. In the course of every modern battle of material a time comes with almost mathematical certainty when the attack, however great its superiority, changes into defence. The course of an entire war under modern conditions is similar. Attack is the less efficient form of warfare with a positive purpose, and it grows weaker the more it exhausts itself in its endeavour to achieve a positive result. In the end the defence is found to triumph.

During the days of crisis following the 12th of April the possibility of a German break-through strategically and not merely tactically successful was carefully considered in London. On the 18th of April Churchill drew up a strictly confidential hypothetical outline addressed to the War Cabinet and discussing the best course to be adopted in the case of a German break-through. The question was whether to give up the right flank or the left; the movement of reinforcements and of ammunition and the distribution of depôts depended on the answer of this question. If the Germans succeeded in forcing a break-through along the coast, the British would be compelled to evacuate the Channel ports one after the other, in which case it would obviously be necessary to maintain contact with the French. On the other hand, if the Germans succeeded in driving a wedge between French and British, the latter could take up a defensive position in front of Calais and Boulogne. In dealing with these possibilities Churchill recommended the adoption of whatever course would be most unpleasant for the Germans. From this point of view there could not be much doubt which was the right answer, and accordingly Churchill recommended the surrender of the coast and closer co-operation with the French army. Thus, even if the worst were to happen, the possibility of an effective defensive programme remained. The Germans would command the Straits of Dover, they would be in a position to blockade the port of London, and they would be able to bombard parts of Kent and Sussex; but they would still be faced by a united Franco-British front which would now be considerably shortened and would be by so much the stronger. The war on land would not be ended before these armies were compelled to lay down their arms.

To assist the defence two so-called water lines were specially devised. The first began at Dunkirk, the second at Gravelines, and both continued to St. Omer and St. Venant; they were called water lines because they contained arrangements for artificial inundation. At first Foch was out of sympathy with this hypothetical strategy; eventually, however, he agreed that the Chief of the British General Staff should work out the necessary plans. The important point was that the water lines had actually been prepared with an enormous expenditure of labour and were ready before the crisis for which they had been devised had become acute.

On the German side, too, there is in existence a grandiose plan of defence. It is a part of post-war literature, and is to be found in Hans von Hentig's *Psychologische Strategie des Grossen Krieges*. But the disciples of Schlieffen, of whom Ludendorff was one, were too completely hypnotised by the idea of attack to be capable of adopting a strategy of defence. The enemy's losses during the great battles from the 21st of March to the end of April were as follows: officers: killed, 2161; wounded, 8619; taken prisoner, 4023. Other ranks: killed, 25,967; wounded, 172,719; taken prisoner, 89,380. Total losses:

officers, 14,803; other ranks, 288,066. The German losses were as follows: officers: killed, 3075; wounded, 9305; taken prisoner, 427. Other ranks: killed, 53,564; wounded, 242,881; taken prisoner, 39,517. Total losses: officers, 12,807; other ranks, 335,962. These figures refute the claims of those who favour attack at all costs on the ground that the defensive is no less costly than the offensive. If we neglect the number of prisoners we find that 308,825 Germans were killed or wounded against 209,466 British. The proportion thus is 3:2.

CHAPTER IX

THE LAST STRAW

Battle of May 1918—Kühlmann's Fall—German War Aims,
July 1918

In spite of the successes achieved by the second great offensive, it was plain to those who had eyes to see that, although ground had been gained, the position of the German army had become less favourable. Looked at from a practical point of view, the salient gained at Merville was a weak spot. It was one more of the "sores" on the German front. The salient was exposed to the flanking fire of the British artillery, and its defence was more costly than its importance warranted. The fighting strength of the German battalions decreased, and it became difficult to make good the losses in killed and wounded, although by now boys of nineteen had been called to the colours.

What inferences did the political leaders of Germany make in these circumstances? What views did they form of the future? Did they possess a plan, and did they undertake anything? There is no answer to these questions; the Government had no plan and did nothing. The only active element was General Headquarters. A few days after the second offensive had been terminated the representative of Headquarters attached to the Foreign Office, Colonel von Haeften, attempted to effect an understanding between Ludendorff and von Kühlmann, the Foreign Secretary. Since Brest-Litovsk relations between these two men had steadily deteriorated, and during the peace negotiations at Bucharest, Herr von Grünau complained to Kühlmann, asking him to explain to the Kaiser that, in view of the continual difficulties caused by the General Staff, it would be necessary in future to hear it only on military questions. In May 1918 an impossible position had been reached: the man who controlled foreign policy and the man who conducted the war had become bitter enemies.

In 1871, after the Crown Prince had made his attempt to effect a reconciliation between Bismarck and Moltke, it was William I. who decided the dispute in favour of Bismarck. In 1918 nothing was done. When Colonel von Haeften tried to heal the breach at the Foreign Office, Kühlmann informed him that it was hopeless to discuss peace conditions with Ludendorff, pointing out that reconciliation was impossible so long as the General opposed an unequivocal surrender of all claims on Belgium. von Haeften refused to take this for an answer, and assured Kühlmann that Ludendorff would raise no objection to any steps initiated by the Foreign Office; but the reply once more was that Ludendorff should apply to Kühlmann personally if he had any political wishes. Ludendorff was not in agreement with von Haeften on all points, and especially did not agree with the latter's optimistic views about the possibilities of peace; there can be no doubt, however, that he enjoyed the full confidence of his chief.

There is no reason for not attaching credence to the account given by von Haeften about his conversation with Ludendorff in the middle of May. According to this version, Ludendorff in his heart of heart had ceased to hope for a favourable issue. "There is no chance of deciding the war in the field unless the combatant forces are reinforced by some 200,000 men of the right quality." According to Haeften's report, Ludendorff went on to say that he had drawn the attention of the Chancellor and the Minister of War to the gravity of the situation. "These remarks of Ludendorff clearly showed me the necessity of bringing the war to a rapid conclusion. Military force no longer sufficed to decide the issue, and it was necessary to have recourse to political and diplomatic means."

Such was the position of affairs at the time of the third great offensive of 1918. Ludendorff was still desirous of carrying out the fundamental idea which had inspired all his campaigns, and which consisted in attacking the British, rolling up their front, and thrusting them into the sea. This plan was impracticable. "To continue the attack near Ypres and Bailleul at a distance of some twenty-five miles from the north coast of France would in itself have been the best plan. The enemy forces on that front were, however, so powerful that this plan was impossible even if our troops had been properly rested. It was necessary for the enemy to become weaker before we could attack upon this front." In order to draw the enemy reserves away from this ideal battlefield, Ludendorff gave orders for an attack upon the Chemin des Dames, a ridge north of the Aisne, west of Rheims, and east of Soissons. This feint was to be followed at the earliest possible opportunity by the main attack upon the British front.

The plan now devised by Ludendorff aimed at reaching its end by indirect means. Foch's reserves, on the Somme and the Lys, prevented him from pursuing his strategic aim, while he was actually embarrassed by his territorial gains. The two salients which he had formed could only be held by large forces of artillery and of infantry which remained tied down in the line. At the same time, in order to prepare the main attack upon the British front, an adequate number of divisions had to be held in readiness behind Crown Prince Rupprecht's lines, so that it became doubtful whether he would have sufficient forces at his disposal for the feint as well as for the main battle. The preliminary staff-work on the attack upon the British front made it clear that the objectives would have to be more modest, and that the action would have to be divided into a number of consecutive parts. On this subject differences of opinion arose; the General Staff of Crown Prince Rupprecht's Army Group, as well as that of the 4th Army, recommended an attack on a wide front with a certain liberty of operations, while Ludendorff favoured moderation. "I think it essential", he wrote to the various commanders, "to clarify views, otherwise I fear that there will be disappointments and heavy losses. We are too short of men to be able to afford the latter."

The greatest drawback to the plan was that it required too much time. The attack upon the Chemin des Dames could not be undertaken before the end of May, and the main attack against the British could not begin before the second half of June. Further, surprise was essential for the success of the plan. No less than 30 divisions and 1158 batteries belonging to the 7th and the 1st Armies, which in turn belonged to the Army Group of the German Crown Prince, were held in readiness on a front of thirty-five miles, ready to commence the attack on the morning of the 27th of May.

On the morning of the previous day two German soldiers, a prospective officer and a private, were taken prisoner by the French. The latter, frightened and confused under cross-examination, admitted that a German attack was impending and that cartridges and hand-grenades had already been issued. The prospective officer, in accordance with instructions, denied everything, whereupon he was informed that, while he could not have been compelled to make a statement, he would be treated as a spy if the information given should turn out to be incorrect. Under this threat he too admitted the truth. By this time it was four in the afternoon and too late for the French General Staff to make any use of the information.

At one in the morning the 1158 German batteries began to fire. The attack was delivered on one of the weakest points on the enemy front, a sector held by 6 French and 3 British divisions. The Chemin des Dames was stormed at the first assault, the Aisne was crossed, and by evening the Vesle was reached. On this first day the German troops had advanced more than twelve miles. The success went beyond Ludendorff's expectations. "I had believed", he says "that we would succeed merely in reaching the region of Soissons and Fismes." From now onwards the battle developed on the same lines as the March offensive; its actual course upset all plans, and the question for the General was whether it would be wise to hold back the advancing centre merely because his strategic plan demanded it.

On the 28th of May the German troops crossed the Vesle. Ludendorff's aims now became more ambitious. On the 29th Soissons was captured, and Ludendorff gave orders for an advance in the direction of Compiègne, Dormans and Epernay. On the 30th he gave orders to surround Rheims on both sides and thus to capture the city. But while his centre advanced as far as the Marne and Château Thierry, forty-five miles from Paris, the wings lagged behind: Ludendorff found himself unable to penetrate into the forest of Rheims or into that of Villers Cotterêts, and on the 5th of June the attack ceased.

When drawing up his battle plans Ludendorff had been compelled to divide the May attack into two parts. The one which had just ended was to be followed at the earliest opportunity by an advance of the 18th Army. The transfer of the heavy artillery from the one front to the other took more than a week, but on the 9th of June the second part of the operation, the battle of Noyon, could begin. This engagement only lasted two days. The infantry of the 18th Army advanced nine miles and captured the heights of Compiègne; on the second day, however, it became apparent that on this occasion the French were prepared. Heavy counter-attacks put an end to the German assault. "The troops concentrated by the enemy", Ludendorff writes, "were so numerous as to induce General Headquarters to suspend operations by the 11th of June. . . . By the middle of June all was quiet on the new front held by the Army Group commanded by the German Crown Prince. There was tension only between the Aisne and the forest of Villers Cotterêts."

There followed one of those pauses between battles which invite reflection. For four months gun, rifles, and *Minenwerfers* alone had spoken; the communiqués of General Headquarters on the results of the latest battles estimated the ground gained and the number of prisoners; but they preserved silence about the reality hidden behind these apparent victories. On the 1st of June, Count Hertling received a letter from Crown Prince Rupprecht which was more eloquent than the official commentaries. The Crown Prince, who commanded on the decisive front facing the British, expressed his conviction that victory was impossible. The issue would be decided by the greater power of resistance and by the greater number of men; and in both respects the enemy held the advantage. "General Ludendorff", Rupprecht wrote, "shares my view that in all probability a crushing defeat of the enemy is out of the question; he is now resting his hopes upon the succour of a *deus ex machina* in the shape of an internal collapse in the Western Powers." Rupprecht did not share these hopes, and pressed on the Chancellor the necessity of taking steps to obtain peace. "I would not have written unless matters were urgent; every week is precious." The Crown Prince further suggested to the Chancellor the possibility of passing on this letter to the Kaiser. Count Hertling apparently failed to understand this appeal, inspired though it was by extreme apprehension; he replied dilatorily and in terms of which only a person completely devoid of judgment would make use. His vocabulary is that of a censored Press; he expresses a conviction that "some more blows of this kind, destructive as they are for the military power of France and England, will end in provoking the populations to a vigorous movement of revolt against their Governments".

At the same time Ludendorff's political right hand, Colonel von Haeften, had drawn up a memorandum recommending a "political offensive" on a grand scale. We saw above that Colonel von Haeften had come to the conclusion that a victory could not be won by arms alone; in his opinion a "political victory" behind the enemy front was also needed. von Haeften's memorandum is notable because it contains a political idea; in those days of complete political sterility, the only person who succeeded in evolving an idea was a soldier. At the same time the idea, as might be expected in an officer unversed in politics, was remarkably naïve, consisting as it did in making the peace of Brest-Litovsk palatable to the British by assigning to Germany the part of champion against Bolshevism. These attempts to exploit any pro-compromise tendencies existing in England were to be supported simultaneously by a public movement in Germany aiming at the support of the movement in favour of peace carried on by Lansdowne in England. von Haeften's suggestion was to collect a number of politicians, economists and scientists, nationalists if possible, and in any case free of any taint of pacifism, who were to agitate in public for his idea. This college of propagandists was to create the impression that the German acquisitions in the East were made in the service of Europe and of culture in general; at the same time "the object would be to inspire the world at large with a feeling of the justice of the German aims, and to convince it that an influential and leading group of men in Germany are honestly combining cosmopolitan and humanitarian ideals with genuine nationalism, far more so than the spokesmen of the Allied Powers". According to von Haeften's plan, the entire demonstration was to be under strict Government control, although appearing as a free and spontaneous movement. It is the misfortune of the military expert that he underestimates the intelligence of enemy politicians as much as he overestimates the talents of his own statesmen. Apart from the fact that such a movement, if it was to be effective, could not be brought about as a command performance, Germany simply was lacking in men capable of casting an ambitious programme in rhetorical form. The politicians were silent, the industrialists pursued concrete gains and annexationist dreams, and most of the scholars of Germany had lost the capacity of forming an independent judgment on questions of the moment.

Ludendorff agreed to von Haeften's plans. He recommended the memorandum to the attention of the Chancellor, observing that it was necessary to exploit the intervals between battles politically, and pointing out that such intervals were the appropriate moment for deliberate political action. There followed a discussion in the Foreign Office, attended by the Chancellor and Colonel von Haeften as well as by Herr von Payer, the Vice-Chancellor; Herr von Kühlmann, the Foreign Secretary; and by his assistant, Herr von Radowitz, during which Hertling and Kühlmann touched upon the weak spot of the plan. They recognised that unless Belgium were surrendered it was useless to hope for any signs of a compromise from England. von Haeften attempted to meet these objections by pointing out that, though he was not empowered by Ludendorff to mention this point, the General's adoption of a political offensive implied that he had adopted moderate views upon the fate of Belgium. Colonel Schwertfeger, in his record of this meeting, drawn up from the official documents, remarks: "All the gentlemen present, especially Herr von Payer, expressed their satisfaction, and considered a peace offensive as practicable and hopeful in the circumstances".

Eventually, however, matters took a different course. von Haeften's memorandum had been passed on to Herr Deutelmoser, the head of the Press department, who had been requested to give an opinion upon it, and who accordingly studied it in the hopes that after all something practical might be made of the plan. He formed the opinion that any political drive must start from the Reichstag. "The Reichstag has from time to time been subject to nervous fits, and for that very reason it is the Reichstag and not the Press, which is, of course, subject to censorship, that is looked upon abroad as the barometer of German public opinion. . . . The sound nationalists referred to in the memorandum are particularly numerous in the Reichstag. . . . If the Reichstag fails to assist us, still more if it should prove refractory, it will be impossible to do anything to produce a good impression either at home or abroad." The chief inference Deutelmoser drew from the memorandum was that apparently the author "did not believe in the possibility of concluding the war by an appeal to mere force of arms". The Press, too, was roused to energy, and the *Kreuzzeitung* and *Kölnische Zeitung* criticised the lethargy of the political leaders in unusually vigorous terms. Finally Herr Helfferich, returning from a visit to General Headquarters, brought with him the impression that "the High Command had obviously given up hopes of finishing the war by exclusively military measures". Accordingly when von Kühlmann took the opportunity offered by a ways and means discussion in the Foreign Office for advocating a peace offensive he was acting in good faith, and in fact his remarks were no more than obvious common sense. "So long as any declaration made by one side is taken by the other as being a peace *offensive*, in other words, as a trap designed merely to mislead and to sow dissension between Allies, so long every attempt at a *rapprochement* will inevitably be denounced. Thus it would appear impossible ever to enter upon negotiations. . . . Without preliminary negotiations, however, the very magnitude of the present war, waged as it is by two sets of Allied nations . . . precludes the possibility of reaching a decision by military operations."

Kühlmann's words were perfectly correct, but the manner in which he blurted out facts had the effect of a bolt from the blue. Between the lines it seemed plain that General Headquarters had ceased to hope for victory. Such, however, was not the case; although Ludendorff and his staff saw that the chances of a decisive battle of annihilation were vanishing, they continued to hold the belief that it was still possible to induce a readiness for peace in the enemy by a series of blows of the kind which had been delivered during the past months. The effect of such operations was to be reinforced, in their opinion, by political action and propaganda for the benefit of the war-weary populations of the enemy countries.

The political offensive devised by Colonel von Haeften was inspired by similar calculations. There can be no doubt that the belief cherished by Ludendorff and his staff was belated; the time had gone when a strategy of attrition combined with a policy of reconciliation could hope to meet with success. But however this might be, Kühlmann's speech was a blunder, and its ultimate aim, which was to induce the Reichstag majority to take charge, was even more inapposite. After Count Westarp's protest nobody dared to support Kühlmann; on the following day the Chancellor disavowed Kühlmann and handed him the knife with which to commit political hara-kiri. His second speech was a pitiful attempt at self-defence; on the 8th of July, Kühlmann left office.

Superficially it might have appeared that he had been ousted by General Headquarters. In the present instance it is indifferent whether the telegram signed by Hindenburg and addressed to the Chancellor was drawn up by the Field Marshal or by his Quartermaster-General; in any case, the "extreme indignation" and "profoundly depressing effect" were heavier metal than Kühlmann, or a better man than Kühlmann, could withstand. Yet even now he might have been saved if the Reichstag had possessed a trace of political instinct and courage, in which case the whole unfortunate affair might, after all, have been of some use. But the Reichstag did not understand the real point at issue, and even Conrad

Haussmann displayed a disarming naïveté at this important political juncture. On the 26th of June, after Kühlmann's speech, he noted in his diary: "Last evening I called on Kühlmann and tactfully told him that it was rash to deliver a speech of this kind without having anything in black and white; his ideas in themselves were correct enough, but a few optimistic passages should have been inserted here and there for tactical purposes". The cautious considerations indulged in by the excellent Haussmann were extremely irrelevant, and Kühlmann might have saved himself in spite of his tactical mistake. "After all," Professor Bredt remarks in his expert opinion, "Kühlmann had adopted the peace resolution of the 19th of July 1917 in its entirety. How was it that in July 1918 not a man was to be found to champion the resolution? . . . It is an incredible fact that the majority parties did not take one more vigorous step in support of the peace resolution. As it was, they allowed Count Westarp to despatch Kühlmann without reflecting that by doing so they sanctioned the victory of the idea of a dictated peace over that of a peace by negotiation."

On the 1st of July Count Hertling, the Chancellor and first civilian in the Empire, practically received orders to report at General Headquarters in Spa: the natural and logical conclusion of the foregoing episode. The minutes of the meeting have been preserved. Dr. Arthur Rosenberg, the Secretary of the Parliamentary Committee of Investigation, describes them as an amazing document.

Chancellor: "von Kühlmann's speech was not delivered in happy circumstances. He was tired, and had not had time to prepare himself. . . . He had not even had time for breakfast, which accounted for his weak delivery and the impression that he was losing the thread of the argument in the second part of his speech. . . . The article by Georg Bernhard also contributed to confuse the Foreign Secretary."

Hindenburg: "General Headquarters have never tried to disguise their suspicions of the Foreign Secretary. The effect of his speech is profoundly depressing. . . . Kühlmann is being backed by the *Frankfurter Zeitung* and *Berliner Tageblatt*."

Hertling apologised to the Generals for the speech of his

Minister "like a teacher trying to excuse a bad essay by one of his pupils before a school inspector."

Another reason why the meeting of the 1st of July was interesting is that on this occasion the Kaiser, the Chancellor, and the Commanders-in-Chief now formed a definite view on the policy to be pursued. The decision arrived at has been laid down by Count Hertling in a memorandum which he left to his son. The historians who deplore the misfortune of Germany during the Great War in being deprived of the "harmonious triumvirate" between King, Statesman, and General, which had existed in the days of William I. are refuted by the minutes of the 1st of July 1918. On this occasion too, Kaiser, Chancellor, and General were unanimous. The three men retained their views that it must continue to be the object of the war "to exert full pressure against England". Thus Ludendorff's strategy of annihilation had the backing of the political leaders. To this programme was added the declaration that Germany was ready "to hear England if approached by the latter, with the proviso that all steps taken must be of the vigour and dignity due to the German effort."

This manifesto by the intelligence behind the weapon, to use Clausewitz's terminology, has been criticised by post-war historians as being based on ignorance of the facts. There may be some justification for this criticism; but it is unfair to describe the manifesto as the outcome of Ludendorff's annexationist dreams. It was, in fact, the manifestation of the intelligence guiding German military policy, and accurately reflected its will and power of judgment. Further, it is in vain to ask which member of the triumvirate carried most weight. Monarchists tend to place the greatest responsibility upon the King and to deplore the unfortunate fact that the monarch did not possess greater insight.

These critics in turn are criticised by Delbrück, who urges that the King must be taken as Nature made him and that the responsibility attaches to his servants. Delbrück considers it not merely "unmonarchistic" but also historically untenable to ask more of the King than he could give. And indeed an unbiased historian will not undertake the attempt to exonerate Ludendorff at the expense of William II., although he will equally reject the converse method and decline to accept a formula which appeals to Gneisenau and Hardenburg, Moltke and Bismarck, in order to justify itself in demanding of the servant the intelligence which the master lacks. Gneisenau and Hardenburg, Moltke and Bismarck, were a part of their time no less than William II., and their intelligence constitutes a collective contribution of the same nature as that of William II. and Ludendorff. To deny that Ludendorff was typical of his age, and to assert that he was a man of doom for Germany, amounts to a distortion of the historical facts, inspired by the bitterness of defeat. The industrialised Germany of William II. did contain a social opposition, but it contained no political intelligence such as had been produced by romantic Germany. What was characteristic of modern Germany was its worship of experts in general, and military experts in particular, the conscious strength of the wealthy classes, and the energy of an entire nation trained in the ways of obedience; but although this was much it was not enough. More than this, however, the joint intelligence of the triumvirate of King, Statesman, and General could not produce.

The remaining points of the German programme were settled at the subsequent meetings of the Crown Council at Spa on the 2nd and 3rd of July. The Austro-Polish solution was abandoned, and Poland was now required to surrender her frontier zone and to pay a contribution to the expenses of the war. Belgium "must remain under German influence to prevent a hostile invasion from ever advancing through it again". It was to be divided into two separate states, to be called Flanders and Walloonia, which were to be united solely through the person of the ruler. Further, Germany required a lengthy occupation, the Flanders coast and Liége being the last points to be evacuated. The final evacuation was to depend upon the "close adhesion" of Belgium to Germany. The Admiralty further required a new neutral zone on the east coast of North America, which, however, the Kaiser declined to grant them. Here again the Kaiser, the Chancellor, and the Generals were unanimous.

The resignation of Kühlmann and the appointment of the

German Minister at Christiania, Herr von Hintze, as his successor, had roused the suspicions of the Reichstag. The Chancellor, however, set these suspicions at rest by assuring honourable members that there could be no question of any change in foreign policy. On the 12th of July he returned to Spa and was able to report to the Kaiser that the Kühlmann crisis had come to a peaceful end. Three days later the Kaiser went to the front. He was full of confidence, and hoped for an overwhelming success.

CHAPTER X

DEFEAT

Third Battle of the Marne—The Black 8th of August—Failure of the Politicians—The Future of Belgium

THE battle which began at dawn on the 15th of July was planned on the same lines as the May offensive. Once again Ludendorff aimed at effecting a surprise attack with superior forces against a weak point of the enemy, breaking through his front and threatening Paris. Since Paris was no more than fifty-five miles distant from the German front, it was hoped to draw the French reserves from the northern front for the defence of the capital, whereupon it was intended to deliver the blow already sanctioned by the Crown Council and prepared some time in advance. This blow was to consist in an attack upon the British by Crown Prince Rupprecht's army. Thus once again the battle would be a feint intended to compel the enemy to weaken his northern front. Ludendorff had decided upon the front on either side of Rheims, since the line from Château Thierry to Verdun appeared to be weakly held.

According to the plan of battle the attack was to eliminate Rheims and the hilly and wooded country lying in its rear on east and west; this difficult sector was to be cut out altogether. Between Château Thierry and Rheims the 7th Army was to cross the Marne and to advance on Epernay on either bank of the river; from east of Rheims to Tahure the 1st and 3rd Armies were to advance, their right wing passing the forest of Rheims, the centre attacking Epernay and Châlons-sur-Marne; and the two inner wings of the two armies were to effect a junction at Epernay. The attack in this instance was on a very wide front; including the sector where no attack

DEFEAT

was to be delivered, its length was seventy-five miles. The forces at Ludendorff's disposal consisted of 47 divisions and over 2000 batteries. The idea of a convergent advance of the three armies was a hopeful element in the plan of battle. At the same time, great difficulties remained to be overcome.

The crossing of the Marne made unusual demands on the troops. Yet it was precisely this part of the operations which was carried through most successfully.

"Down from beyond the German parapets", so Churchill writes, "leaped the cataracts of fire and steel. Forward the indomitable veterans of the Fatherland! It is the Marne that must be crossed. Thousands of cannon and machine guns lash its waters into foam. But the shock troops go forward, war-worn, war-hardened, and once again 'Nach Paris' is on their lips. Launching frail pontoons and rafts in a whistling, screaming, crashing hell, they cross the river, mount the further bank, grapple with the French; grapple also with the Americans—numerous, fresh, and coolly handled. After heavy losses they drive them back, and make good their lodgments. They throw their bridges, drag across guns and shells, and when night falls upon the bloody field, 50,000 Germans have dug themselves in on a broad front four miles beyond the Marne. Here they stop to gather further strength after performing all that soldiers have ever done."

Ludendorff, more sparing than Churchill in his praise of the military performance of the German troops, described the crossing of the Marne as a remarkable achievement.

Three miles south of the Marne the enemy resistance became more stubborn, and it was found impossible to compel him to withdraw further without throwing considerable artillery forces across the river. The fight came to a standstill. On the front of the 1st and the 3rd Army the enemy retreated according to plan into their second lines, whereupon the attack came to an end on this position too. "By noon of the 16th of July," Ludendorff writes, "General Headquarters gave orders to the 1st and 3rd Army to cease attacking and to take up defensive positions. To continue the attack would have proved too costly. . . . Once I had reluctantly come to this decision a further advance beyond the Marne became unnecessary. There was nothing to be gained by leaving our troops on the southern bank . . . and it became necessary to arrange for a retreat. . . . On the 17th of July, orders were given to withdraw. The troops south of the Marne had a difficult time; their behaviour was extremely praiseworthy."

It is worth while to draw attention to the heroism of the German troops in this battle, since this was the first engagement which did not meet with even a tactical success. "For the first time", General von Kuhl says in his expert report, "the German attack had proved a failure tactically . . . the methods which had proved so successful hitherto had proved unavailing."

What had happened? The enemy had learnt the time and place of the impending attack. Churchill says that the information of deserters and prisoners made in special raids had provided British and French Headquarters with consistent information allowing the necessary preparation to be made in good time.

The French Intelligence Service of Belfort, too, was, according to Kuhl, informed from Swiss sources of the coming offensive. "While the Staff was at pains to preserve the utmost secrecy", Ludendorff writes, "the innate loquacity and boastfulness of the Germans brought our most secret plans before the public, and thus before the enemy." The publication of the German intentions enabled General Gouraud to forestall the preliminary bombardment of the 1st and 3rd German Armies and to allow the German attack to exhaust itself upon a skeleton front while the second French line, which was the one held by the bulk of the French troops, lay beyond the range of the German artillery. In this way the German attack lost most of its force and came to a standstill before the real objective was reached. The decisive events, however, came later and had nothing to do with the fact that the German intentions had been revealed beforehand.

Having given instructions for the retreat Ludendorff reverted to his main idea, namely, the Battle in Flanders, and this despite the fact that he had failed in his intention of weakening the enemy front in that region. In the night of the 17th-18th July, Ludendorff proceeded to Crown Prince Rupprecht's headquarters at Tournai. It was there that he learnt on the following morning that a sudden tank attack, delivered by the French south-west of Soissons had broken through the German front. He concluded his discussions at Tournai and returned to Avesnes in a state of the "greatest anxiety". This was Foch's first success. It had been prepared since June, and Foch had refused to give it up in the face of every kind of resistance.

While Foch was Commander-in-Chief of the French and British forces, he was not chief of the French General Staff. He had a small personal staff, his "family", but the "Grand Quartier Général" was Pétain's preserve. The latter was more concerned with the safety of Paris than with any ideas of an offensive, while the British Government became anxious when Foch, on two separate occasions, insisted on withdrawing four British divisions from the Flanders front. This was the critical moment in Foch's career. He refused to allow himself to be flurried even when he learnt of Ludendorff's intentions.

Ignoring the German threats, Foch concentrated 20 divisions in the extensive woods round Villers-Cotterêts. All movements took place at night; by day all troops, transports, and trains remained in their positions, except when they were covered by the woods. On the morning of the 18th, the French emerged from the forest. There was no artillery preparation and a way was cleared for the infantry by 330 small Renault tanks. The French advanced for a distance of 3 miles on a front of 28 miles. After the last German lines had been passed, the French came upon German soldiers who were quietly cutting corn. "The Germans threw their scythes away and fought where they stood."

The ensuing battle embraced the entire salient between Soissons and Rheims. From the 18th of July to the 2nd of August, a difficult and costly retiring action was fought, during which the salient captured in May was gradually evacuated in its entire depth of 18 miles. By the 2nd of August the German front was again behind the Vesle, in other words, north of the straight line between Soissons and Rheims. The remnants of 10 divisions had to be distributed among other units, and these heavy losses compelled Ludendorff to give up the idea of the main offensive in Flanders. "Undoubtedly we had suffered a severe reverse," was the judgment of General von Kuhl. This was not the fault of the troops: "the conduct of the men had been admirable and the retreat had been carried through in good order." Ludendorff, too, admits that "throughout the battle officers and men did the utmost that could be asked of them. . . . Great as had been the demands made of the troops, our men had offered a vigorous resistance and felt superior to the enemy."

Innumerable critics have examined this so-called turning point of the war with a view to discovering the source of our failure and determining where the fault lay. In Stegemann's opinion the intervals between the various battles beginning in March 1918 were too long. "These intervals were made necessary by the fact that the Germans had to observe a certain economy in utilising their forces, so that the only recourse possible was to discontinue the offensive." In other words, Germany was too weak to deliver her blows in such rapid succession as to make it impossible for the enemy to bring his reserves to the threatened spot. Hence Stegemann infers that the only possible course was to give up the offensive. Such a method, however, was alien to Ludendorff's ways of thought and indeed to those of the Prussian General Staff as a whole, as appears from the reply given by General von Kuhl to Stegemann's criticism:

"In what manner was the campaign to be carried on? To give up the offensive implied a return to the defensive, and a return to the defensive was tantamount to losing the war. The only justification for carrying on was the hope of obtaining acceptable terms at the end. There was, however, no cause for such a pusillanimous resolve so long as it was possible to make a last attempt to turn the war in our favour." This reply agrees exactly with Ludendorff's ideas and with the aims and views of the General Staff. The aim of the war as envisaged by the German military leaders did not consist in "acceptable conditions"; a peace without annexations was not their aim. Ludendorff himself has said "victory or defeat; there was no third way." His critics, led by Delbrück, exhaust themselves in attempts to demonstrate that there was a third possibility, consisting in a peace without annexation. Such endeavours are theoretical exercises and have nothing to do with reality; the war aims which Germany pursued in July 1918 could not be reached otherwise than by an overwhelming victory—in other words, by the annihilation of the enemy; and consequently in pursuing the annihilation of the enemy Ludendorff was consistently carrying out the German programme. The problem of Ludendorff lies elsewhere: in the fact that he did not grasp the nature of a modern battle.

It is impossible to form a correct idea of the operations beginning with the March offensive so long as they are looked upon retrospectively as a connected series of operations directed to one common end. Such was not the case. Ludendorff based his strategy upon the belief that a strategic breakthrough was possible, and even after the failure before Amiens he did not see the error of his ways. We have here the line of demarcation between the soldier of genius and the expert. A real soldier possesses intuitive powers and is capable of appreciating innovations: the expert is the prize boy of a military academy, and always remains tied in academic bonds. Throughout 1918 Ludendorff pursued the phantom of a battle of annihilation. On two occasions he came near to effecting a strategic break-through, and it was precisely because he came so near to success that he felt encouraged to attempt a third offensive on these proved lines. In fact the German successes did more harm than good because they left the German position worse than it had been at the beginning. Foch has described this method as a bull-like strategy; but this description is merely offensive and explains nothing, and Foch himself never suspected that these elaborate attempts at a break-through with all the sacrifices that they entailed implied the end of a great military tradition. Churchill, casting about for an explanation of Ludendorff's conduct, translates the military expert into a more familiar sphere by saying that the German general was a desperate gambler incapable of giving up the game as long as he had a stake to venture: who can say whether at this moment he was right or wrong?

The Allied generals had realised that a strategic breakthrough was impossible. They put their faith in the pressure exerted by superior numbers, in the "Iron Infantry" of the tanks and in the superiority of men and material at their disposal. Their intention was to break the enemy, not by a battle of annihilation, but by attrition.

By the beginning of August the whole Western front was on the defensive. The last battle had seriously weakened the German forces, and it became necessary to bring them up to strength by drafts upon the reserves which it became increasingly difficult to make. The divisions which had been detailed for the Flanders attack had been used up partly during the last battle and partly during the defensive actions; in Crown Prince Rupprecht's Army Group alone the daily losses amounted to 1200 men. At the same time the whole front was being ravaged by Spanish influenza. By the middle of July the average strength of a battalion in Crown Prince Rupprecht's Army Group was 670 men: Ludendorff was compelled to prepare his front for the defensive. "Further, it was necessary to arrive at a decisive resolution in consultation with the Chancellor, who was kept permanently informed of events. . . . While I was considering these steps the blow of the 8th of August was delivered. . . . In the annals of this war, the 8th of August 1918 is the black day of the German Army."

On the 24th of July Foch had explained his plans to Generals Haig, Pétain, and Pershing. These plans covered the campaign of 1918 and 1919, and envisaged separate attacks on the three German salients, with the idea of improving the railway connections from the sea to the Vosges, to be followed by a general offensive. At British Headquarters scepticism prevailed, as the last French offensive, like all other offensives of the old style, had met with tactical successes, but had failed to pierce the enemy front. Eventually, however, Haig was persuaded to detail the 4th Army under General Rawlinson for an attack on the German salient before Amiens. This operation was to be a tank offensive; 600

DEFEAT

tanks were to deliver a surprise attack under cover of a smoke screen and without artillery preparation. The British attack was to be supported by a French force under General Debeney.

A little after 4 on the morning of the 8th of August the 600 tanks moved against the German front between Albert and Moreuil, followed by 4 Canadian, 4 Australian, and 2 British divisions. The following is Ludendorff's description: "They succeeded in advancing for a considerable depth beyond our front. The divisions facing them allowed themselves to be completely overrun. Divisional staffs were surprised by the tanks in their Headquarters. . . . In the early hours of this day I fully appreciated the position; it was a very melancholy one . . . six or seven German divisions which could only be described as perfectly fit for battle, were in a state of complete dissolution. During the first two hours the British took 16,000 prisoners and captured 200 guns." The German retreat lasted until the 12th of August. Local engagements now began to take place, and, as Ludendorff records: "Our troops again stood fast". The German losses, however, had been very severe.

This battle resulted in a considerable tactical success of the enemy; at one point the advance amounted to twelve miles; the enemy had taken numbers of prisoners and had captured a good deal of material. On the other hand, although successful, this operation did not differ essentially from the last German offensive; indeed, as far as the gain of ground, the number of prisoners, and the material captured was concerned, it was less successful than the German operations. The Allies had not succeeded in effecting a strategic break-through any more than the Germans; both sides had proved unable to roll up the enemy front. Why then did Ludendorff refer to the 8th of August as the black day of the German Army? His own explanation is that it was apparent for the first time during this battle that the German Army was breaking up from within. "Acts of the utmost courage were reported, but I also heard of things which I should have believed impossible in the German Army, troops surrendering to single cavalrymen, and entire companies to tanks. The men of a division who went up to the attack resolutely were called black-legs by retiring troops. . . . Despite repeated warnings, all my worst fears were here realised simultaneously; the German Army had ceased to be a perfect fighting instrument. . . . To carry on the war became a gamble for which I could not accept responsibility. . . . The fate of the nation was too high a stake for a game of chance. The time had come to end the war."

In itself Ludendorff's version was correct; while there were instances of the utmost gallantry and devotion, things happened on the 8th of August which had hitherto been unknown in the German Army. But Ludendorff's explanation does not go to the root of the matter. We have remarked on a previous occasion that it is impossible in any case to cast doubts upon the fighting spirit of an army which had been engaged upon offensive operations for four years; and the same applies to the 8th of August. Since Napoleon no general enjoyed greater confidence than Ludendorff; under-fed, ill-equipped, and driven without rest from battle to battle, the German infantry yet responded to every call. For four years nothing had been able to break the spirit of the German Army; its courage had made good the enemy's superiority in numbers and ammunition no less than the defects of the German military machine. The men trusted their leaders, and their confidence was such that it survived every defeat; "we retired singing to the Siegfried line", Hans von Hentig writes, "confident that we had established superiority over the enemy by an intellectual operation".

The so-called failure of the 8th of August was symptomatic of something more than is implied in the simple explanation which accounts for every failure by the alleged undermining of discipline and of the spirit of the offensive by the machinations of agitators. Instances of an ostentatious refusal to comply with orders were the more striking because they were rare, and left no room for doubt that the enormous moral credit which General Headquarters had enjoyed was exhausted. The events of the 8th of August amounted to a vote of no-confidence passed by the German Army upon its leaders. Ludendorff himself was well aware that the trust of the Army was on the wane; but he felt himself responsible only to the Kaiser: "I thought it possible that these events might have shaken His Majesty's confidence in me, and there was a chance that a new commander might be capable of a more unbiased judgment. I therefore deliberately suggested to General Hindenburg the advisability of finding another man to fill my place. The Field-Marshal declined. I also spoke to the Head of the Military Cabinet, asking him to find a substitute for me if any doubt was felt as to my capability. However, the Kaiser showed me special marks of confidence during these days." After the discussion at General Headquarters at Avesnes, the Kaiser summed up as follows: "I see that it is necessary to review the situation. We have come to the limit. The war must be brought to an end. . . . Accordingly I shall expect the Commanders-in-Chief at Spa in the course of the next few days." This, as Lieutenant-Colonel Niemann says, was "the decisive word, which Ludendorff and Hindenburg, whose function it was to fight while there was a will to fight, had hesitated to say."

There are few events in history which have been studied as thoroughly in all their details as the discussions which took place at General Headquarters at Spa on the 13th and 14th of August 1918.

The two Generals arrived at Spa at 8 in the morning on the 13th of August. A meeting had been arranged at 10 o'clock between them, Count Hertling, the Chancellor, and Herr von Hintze, the Foreign Secretary. Before the meeting, however, Ludendorff took Herr von Hintze aside and told him that while he still had felt confident in the middle of July to break the resolution of the enemy by his offensive, and thus to compel him to sue for peace, he now no longer held this belief. On being asked how he intended to continue the war, the General replied that a strategic defence would eventually succeed in wearing out the enemy. This description of the military position was given in confidence, and, as Hintze himself says, completely overwhelmed him.—At that time Ludendorff and Hintze were on completely confidential terms; Hintze was "the chosen man of Headquarters".

After the confidential discussion between Ludendorff and

Hintze the meeting between Hertling, Hintze, Hindenburg, and Ludendorff took place. On this occasion the two Generals confined themselves to a brief description of the military position, pointing out that while an offensive on a great scale was no longer possible, occasional minor offensives were still feasible. Both were convinced that this method, although on the whole of a defensive nature, would succeed in making the enemy ready for peace. Having made this declaration, the Generals turned to another subject. They complained that a bad spirit was abroad in the country, that the authority of the State was being weakened and that an unsatisfactory reaction in the Army was inevitable. These charges were followed by a lengthy debate. Hintze attempted to concentrate the discussion upon those subjects which, as he said "required a new definition in view of the necessity of diplomatic peace overtures", in other words, on Belgium and Poland. But as soon as the word "Belgium" was mentioned Ludendorff exclaimed: "Why bring up Belgium? That question has been settled and is laid down in black and white", whereupon a new subject came up for discussion. Hintze confined himself to a description of the political position. His view was gloomy in the extreme, with Austria near to collapse, Bulgaria ready to desert the Central Powers, and Turkey pursuing a policy of her own. Ludendorff described this view as morbidly pessimistic.

On the following day Hintze called upon the Chancellor. The more optimistic official view maintained by the Generals, so far from reassuring him, had only increased his pessimism overnight. He now intended to ask the Kaiser for permission to enter upon peace overtures, and requested the Chancellor to support him, assuring the latter that, if he failed to obtain the Kaiser's sanction, he would be compelled to tender his resignation. Whereupon Count Hertling replied: "I am an old man; let me go first."

There now followed a discussion in the presence of the Kaiser. The meeting was attended by the Crown Prince, the Chancellor, Hindenburg, Ludendorff, von Hintze, von Plessen, Herr von Berg, the Head of the Civil Cabinet, and Freiherr von Marschall, the Head of the Military Cabinet.

The official minutes and Hintze's private notes contain the outlines of the debate, which was opened by the Chancellor. Count Hertling gave a review of the internal position, pointing out that the population was war-weary, food supplies inadequate, that there was a lack of clothing, and that an electoral reform was essential. After him Ludendorff spoke. According to him stricter discipline was wanting; Lichnowsky ought to be locked up, and there ought to be a combing out of all young men not yet called to the colours.

The actual subject for debate was thus not touched on by the Generals. On the other hand, Hintze could not bring forward his wishes without giving an outline of the military position. Accordingly he informed the meeting of what Ludendorff had told him. He pointed out that the Chief of General Staff was of the opinion that the enemy could no longer be beaten by military operations, and that, consequently, it must be the aim of the military leaders to make the enemy ready for peace by a strategic defensive. "The political leaders accept the verdict of the greatest General produced by the war. They infer that we are unable to break the enemy by military action, and that the policy to be pursued must take into account the military position." When the minutes were drawn up, a mistake was made which might be of interest to students of Freud. In the passage where the political leaders make their inference, the original said: "that we are unable to break the enemy by political action. . . ." Count Hertling amended the minutes by writing ". . . by military action?"

After Hintze, the Kaiser spoke and pointed out the necessity of better discipline on the Home front. On other matters he was in agreement with the Foreign Minister. The most important passage of his speech was one where he pointed out the necessity of watching for a suitable opportunity of entering upon negotiations with the enemy. Such an action might be initiated through the intermediacy of the King of Spain or the Queen of the Netherlands. The last word lay with Hindenburg. His observations are summed up in the minutes in the following terms: "General Field-Marshal von Hindenburg expressed the hope that we might succeed in holding French ground and in imposing peace upon the enemy in this way." Ludendorff tells us that he himself substituted "opinion" for "hope" in the minutes, thus introducing a note of greater confidence. After the meeting Hintze was in a state of great emotion: "He had tears in his eyes.... When I shook hands with him I was deeply moved myself."

After the collapse a lengthy controversy centred about these two discussions at Spa. On the one hand von Hintze has attempted to cover the German diplomacy against the charge that its methods in initiating peace negotiations after the events of the 14th of August were unnecessarily slow and clumsy. He puts the blame on the military leaders, and urges that they lacked the courage, even at this critical moment, to describe things as they were. According to him, the discussions at Spa were confused by their unjustifiable optimism, with the result that no definite conclusion was arrived at. Ludendorff, on the other hand, rightly points out that in fact complete unanimity prevailed at Spa. Nobody would have prevented von Hintze from expressing his doubts and from pointing out the necessity of taking immediate steps in order to ensure peace. Thus the dispute really amounts to the question whether the Kaiser would have given orders to take the initial steps for peace negotiations if he had been given a clearer description of the actual military position. The quarrel lies between the Generals, who had a thousand reasons for not proclaiming the full truth, and the diplomatists, who were somewhat hard of hearing. In actual fact, von Hintze must have grasped the reality which was hidden behind the words of the Generals, since, otherwise, it is impossible to account for his pessimism, his tears, and his general excitement. Surely, at that moment pedantic explicitness had become superfluous. It was plain enough that the Kaiser wanted peace, and that his further instructions to the effect that it would be best to await an improvement in the military position did not sanction dilatory methods. It was by no means certain that there would be any improvement, and indeed there was every possibility that it would never come at all. von Hintze was a poor psychologist if he failed to perceive that the Generals could not act otherwise than they did.

Indeed, the present is yet one more instance of the experts' tragedy. They were unable to play the historic part which they were called upon to fill, nor is there any need to follow the example of the Parliamentary Commission of Investigation in attempting to delimit responsibility and to discover the real culprits. The characteristic of this melancholy episode is the complete lack of judgment of the political leaders. There was a blind trust in the military experts, with the result that no politician dreamt of thinking independently and checking facts for himself. As late as July von Hintze hoped that it might be his "pleasant and promising part to crown a certain victory with a victorious peace."

It is absurd to pretend that the position of Germany could change in four weeks from one justifying the brightest optimism to one that called for the most hopeless pessimism. Any diplomat capable of independent judgment would have known a month ago what von Hintze did not know, namely, that Germany was engaged in a desperate struggle, and that a victorious peace was out of the question. It is true that such an opinion could only be reached by a man who possessed not only a knowledge of the facts, but also an intellectual independence which was far from being characteristic of the political leaders of Germany. Optimism was ostentatiously displayed by the military leaders, who saw in it a virtue; but such optimism was not a fair substitute for independent political judgment. The 14th of August threw a cruel light on the blindness of the German diplomats; and Hertling's and von Hintze's method of flight from this dangerous ground was no solution of the difficulty.

It is worth asking whether at this moment any chance remained of initiating peace overtures as envisaged at Spa. In his *Entstehung der deutschen Republik* Dr. Arthur Rosenberg considers that there was only one means of salvation; in his opinion the only salvation from the impending disaster consisted in a radical revolution; the Reichstag majority should have deposed Ludendorff in August and should have set up a parliamentary government in his place. On this somewhat impossible assumption a vigorous revolutionary Government would have arisen which would have wound up all the adventures in the East, transferred the troops to the Western front, brought about a political understanding with Soviet Russia, compelled Austria-Hungary to adopt a Federal Constitution, and have given autonomy to Alsace-Lorraine. "German co-operation with Russia would have brought about a change in the Balkans; and faced with a firm political bloc consisting of Germany, Russia, and Austria, the Allied Powers would have been much readier to think of equitable terms of peace than when dealing with an isolated and defenceless Germany." Such a calculation required political gifts in the Reichstag which did not exist in Germany under the Imperial regime and of which no traces have appeared under the Republic; if Germany had possessed such a gift of political judgment she would never have drifted into the position in which she found herself in August 1918.

As it was, Herr von Payer, the spokesman of the Reichstag, far from taking up the initiative and entering upon an independent policy, could think of nothing better than to apply to Ludendorff with the request that he should moderate the German war aims. The negotiations between Ludendorff and von Payer at Avesnes are rightly described by Rosenberg as a grotesque historical spectacle. The situation, however, is incorrectly described when he goes on to say that Ludendorff, although vanquished in the field, continued to play the dictator and attempted to save the remnants of a victorious peace. It is not Ludendorff but Herr von Payer who appears in a grotesque light when in the name of the Reichstag majority he asked General Headquarters to listen to reason and give up all claims on Belgium.

We know that Ludendorff had drawn up a Belgian programme which implied an overwhelming victory. By the end of August 1918, few things were less important than what the Generals thought about Belgium; "what General Headquarters thought about war aims had ceased to matter; the question now was how long the Kaiser was likely to remain on the throne." In all probability Ludendorff would have been only too glad in those days to give up all claims on political authority. His thoughts were busy with the front; and it required all the characteristic lack of psychological insight of a German politician to fail to grasp the situation. von Payer and von Hintze were so incapable of independent action that they continued to thrust the decision upon the military expert.

Thus forced to take the initiative, Ludendorff continued in his obstinate course. Jointly with von Payer, von Hertling had drawn up a memorandum on Belgium in which Germany declared that she would surrender Belgium "unconditionally and without any claims" on the conclusion of peace. At the end of this document there is a passage declaring that it would be necessary to reach an agreement on details, a clause concealing the German claim to solve the Flemish question in her own interests. But this clause is evidence only of the blindness of its authors; the whole declaration was worthless and it did not matter if Ludendorff added a martial flourish.

Six precious days passed in the drafting of this document; on the 27th of August the final form was drawn up. The declaration began with an attack against England, which claimed to be fighting on behalf of Belgium, but was aiming at the destruction of Germany and a permanent occupation in Belgium and France. In the final sentences it was expressly stated that the Flemish question was one of the details which made it necessary to come to an understanding. von Hintze had at Washington an intermediary for dealing with Belgium who had been instructed on the 14th of August that all questions were opened to discussion on principle; but the final Belgian formula never reached this intermediary. Things continued to grow worse on the German front, and it became superfluous for Belgium to reply to the Hintze overtures. The Belgian Government publicly declined to enter into negotiations, and the Belgian intermediary was instructed not to forward any fresh suggestions.

Rosenberg compares Ludendorff and von Payer in their lengthy discussions with two men playing chess on a sinking ship, who continue to check each others kings while the waves are about to close above their heads. The military expert without political understanding and the politician with his invincible respect for the expert are alike unable to understand the realities of the situation.

CHAPTER XI

COLLAPSE

Lack of Plans for a Defensive—Bulgaria Surrenders—No Hope Left—Constitutional Troubles—Note to President Wilson—Ludendorff a Last Ditcher—Ludendorff's Resignation

On the 28th of August, while tracing the German retreat on his chart, Marshal Foch remarked to his aide-de-camp: "the man could escape even now if he would make up his mind to leave behind his baggage." The great question was what the Germans would now do; whether the German Army would collapse as it did after Jena or whether it would resist to the bitter end. Secretly the enemy hoped that there would be a second Jena; but the battles of the last week of August had been so costly, and the German resistance so stubborn, that common sense did not sanction any hope of a rapid collapse. In Churchill's opinion the best thing for Germany would have been to effect an orderly retreat to the line of Antwerp and the Meuse; after the battle of the 8th of August it was the first duty of the military leaders and statesmen and of every party and class in Germany to reach this line at all costs. The Allied leaders in fact thought such a retreat probable, and feared that the Germans would leave behind shells and land mines fitted with time fuses—a German invention—set to explode weeks or even months later. The Allied Armies would have been unable to carry out a pursuit in a region where railways and roads had been destroyed; and their generals were haunted by the fear of a six months' interval during which the Germans would have been able to select and fortify a new line of defence.

Such military advantages did not, however, weigh as

heavily with the Allied Powers as the moral consequences of the evacuation of France and Belgium. The liberation of French soil was the idea which upheld the French nation and caused it to continue the war, while the liberation of Belgium was still the chief motive why Great Britain did not give up the fight.

"Had Germany, therefore, removed both these motives, had she stood with arms in her hands on the threshold of her own land ready to make a defeated peace, to cede territory, to make reparation; ready also if all negotiation were refused to defend herself to the utmost, and capable of inflicting two million casualties upon the invader, it seemed, and seems, almost certain that she would not have been put to the test. The passion for revenge ran high, and stern was the temper of the Allies; but retribution, however justified, would not in the face of real peace offers have been in itself a sufficient incentive to lead the great war-wearied nations into another year of frightful waste and slaughter. In the lull and chill of the winter, with the proud foe suing for terms and with all his conquests already abandoned, a peace by negotiation was inevitable."

In Churchill's opinion, the motives which prevented German Headquarters from ordering a rapid retreat to the Antwerp—Meuse line were of a thoroughly inadequate nature. The German General Staff could not bear the thought of sacrificing the enormous quantity of ammunition and material of every kind which Germany had accumulated in France and Belgium in the course of the last four years, and which now became a dangerous burden.

It would be doing less than justice to the German General Staff to assume that the only reason which prevented it from deciding upon a general retreat consisted in reluctance to give up so much war material. As a matter of fact, the views of army commanders differed. Thus Gallwitz recommended holding the line, with occasional systematic counter-attacks; while the General Staff of General Boehn and the Crown Prince thought it best to work out a plan of defence against the elaborate attacks to be expected. The Crown Prince himself and Count von der Schulenburg advocated a shortening of the front from purely military reasons, since this was the only way to obtain reserves. "We thought it imperative to retreat immediately to the Antwerp—Meuse line, and further envisaged a retreat on a large scale as far as the line Maastricht—Luxemburg—Metz—Strassburg—Upper Rhine. We fully appreciated the drawbacks involved in the loss of all our stores of war material, but such a loss was preferable to the destruction of the Army."

Ludendorff himself did not approve of the idea of such a general retreat. His first orders for the preparation of a new line were issued to the army groups commanded by the Crown Prince and Crown Prince Rupprecht on the 6th of September. The line in question was not, however, the Antwerp—Meuse line but the so-called Hermann line, which ran east of Bruges and along the Lys and the Schelde, and the Hunding—Brunhilde line which followed the course of the Serre, the Souche, and the Aisne and ended in the Argonne. Ludendorff himself has answered the question why he did not decide in August to retreat to the Antwerp—Meuse line: the fact is that the line existed only on paper.

"We had not the men necessary for building new lines, and this is the simple reason why they were never constructed." To construct a defensive line of some hundreds of miles is an elaborate work requiring a vast number of different operations. The line has to be determined, a plan for occupying it has to be worked out, and engineers have to be detailed, together with the necessary workmen and means of transport. Accommodation has to be found for the men, special railways have to be built, transport to be found, and the building material purchased and despatched. Further, trigonometrical work has to be done to allow maps to be made, hydrographic and geological studies have to be undertaken, the evacuated zones have to be devastated, etc. Between the autumn of 1916 and the spring of 1917, the labour companies of the German Army finished the Siegfried and the Michel line and constructed a second line from the sea to Verdun. After the spring of 1917 their services were required by the great offensives.

The investigators who have made it their business to study

the collapse of Germany have given due weight to all these details; but it would amount to not seeing the wood for trees to overlook the fact that the very nature of the German strategy involved a neglect of the defensive. A grandiose plan of offensive and a grandiose defensive are incompatible. If a plan of campaign aims at a decisive attack culminating in a strategic break-through and the annihilation of the enemy it cannot simultaneously aim at a comprehensive system of defence.

Ludendorff was the pupil of the old Prussian military academy, and as such he was doomed to be consistent to the end. With him the military means and the political ends had one aim, which was to attack and annihilate; and the natural means to this end consisted in conquest. A different method of warfare would have required different aims and a different manner of thought. But such aims and method of thought have never been effective elements in German military history.

The controversy covering the period from the 14th of August to the collapse consists in numerous variations of the question why the admission of defeat took place like a bolt from the blue. The whole problem has been examined in minute detail, and various students have investigated the course of events from day to day and from hour to hour with microscopic attention. The last act of the great drama has been analysed in all its aspects, and the part of every actor can be followed exactly. An enormous mass of events is compressed within a few days; a hundred causal series meet at one point; material and intellectual factors operating in a struggling world culminated in one resultant; and among these factors the blunders of the German political and military leaders were not the least important.

While von Payer was engaged upon working out a formula containing the simultaneous surrender and retention of Belgium, Austria had made up her mind to sue for peace independently. On the 30th of August, Prince Hohenlohe, the Austrian ambassador, conveyed this unexpected news. Three days later the British broke into the Wotan line between Arras and Cambrai. On the following day Count Hertling gave a vague outline of the position in the Prussian Cabinet. The Ministers were at a loss; eventually the Chancellor promised to ask General Headquarters about the military position. Meanwhile von Hintze sent a telegram from Vienna saying that he had succeeded in preventing Austria from making a separate peace, but pointing out that Baron Burian in return requested Berlin to take serious steps to induce the Queen of the Netherlands to act as mediator. Hintze asked for two weeks' grace, and hoped that the front would be secured during this interval.

Within twenty-four hours Austria had changed her opinion. She threatened to publish the so-called "universal appeal", while Baron Burian caused it to be known in Berlin that Austria had come to the end of her tether. General von Cramon, the military plenipotentiary at Vienna, succeeded in inducing the Emperor Charles to grant a brief respite, with which however was coupled the express condition that Austria was to be informed when the German Army had reached its last line of resistance. On the 9th of September von Hintze proceeded to General Headquarters at Spa, carrying with him a list of questions drawn up by Hertling. Ludendorff's summary answer was: "The general idea of the defence is to remain where we are", and it is difficult to say what other answer he could have given to a person so incapable of independent judgment. On the following day this reply was passed on to Vienna. Hintze now had in his possession a draft drawn up by General Headquarters. This document was quite unequivocal. "I cannot agree that Austria is to despatch her 'universal appeal'; such a step would be disastrous. On the other hand I am ready to accept the immediate intervention of a neutral Power." Surely nothing could be more explicit.

On the 11th of September, Hintze telegraphed to the German representatives at Vienna, Sofia, and Constantinople, to the effect that the Kaiser and General Headquarters had given their consent to an immediate *démarche* before the Queen of the Netherlands. On the 12th of September a vigorous American attack west of Pont-à-Mousson compelled the Germans to evacuate the St. Mihiel salient. By now Austria had determined to go her own way, and a telegram from the Kaiser to the Emperor Charles came too late to change this resolution. On Sunday the 15th the Austrian peace note was published. At last the party leaders were startled; everyone spoke of a bolt from the blue, and thereby displayed his own lack of political judgment.

In the forenoon the popular leaders called upon the Chancellor, who attempted successfully to set their minds at rest. At the same time, however, it occurred to them that it might be worth while to ask that the main committee should be convened. On the same day news was received that the Allies had broken through the Bulgarian front between Cerna and Wardar. von Hintze continued his efforts to induce the Netherlands to act as mediators, his idea being to let the Austrian *démarche* be followed immediately by Dutch mediation, a plan which met with very moderate success; the Netherlands merely promised to allow the delegates to assemble at The Hague. The Foreign Office considered the possibility of offering the part of mediator to President Wilson, and on the 21st Ludendorff recommended Prince Hohenlohe-Langenburg for this mission. Bad news was received from Constantinople; the Turks had suffered reverses in Palestine and Mesopotamia.

On Tuesday the 24th the main committee assembled and was addressed by Hertling, who spoke vaguely, and, knowing little himself, was unable to convey much information. On the Thursday von Hintze telegraphed to the German ambassador at Vienna that he feared the secession of Austria at any moment, and that he recommended an occupation of the Danube line in order to "keep Austria up to scratch". Ludendorff sent troops to Bulgaria, which arrived too late, and on Friday, the 27th, Hintze had to inform the main committee that Bulgaria had surrendered. This made a profound impression upon the Reichstag. At the same time General Headquarters sent out a despairing cry; von Lersner, the representative of the Foreign Office, reported to von Hintze that General Bartenwerffer, and Colonels Heye and Mertz, were extremely pessimistic, and that Ludendorff was clinging "like a drowning man" to the report that the French Army was ravaged by pulmonary disease. Ludendorff himself used this expression in the presence of members of his staff.

A commotion among the populace and in the army now seemed imminent, and Hintze began to think of precautionary measures. He suggested the formation of a popular parliamentary Government; whereupon Count Hertling asked him indignantly whether he could dream of having Social Democrats in his Cabinet. von Rosenberg, von Bergen, and von Stumm, of the Foreign Office, worked out a plan for a more parliamentary Government, of which Herr von Payer was to settle the details in consultation with the party leaders: eventually this new Government was to ask President Wilson to undertake the office of mediator. On the morning of the 28th September von Hintze informed the Chancellor that he was on the way to Spa "in order to ask General Headquarters to give him a definite and unequivocal declaration on the military position, and in order to suggest any steps that might become necessary."

On his own showing, Herr von Hintze is proved to be the man who had ceased to hope for victory on the 13th of August. It is difficult to understand the precise nature of the hesitancy which beset him from that day onwards. His idea of things in general was approximately correct, but he had not the courage of his convictions. On each and every occasion he waited for the confirmation of the competent authority, and was not satisfied with the fact that the language of the Generals plainly foretold the worst to anyone who had ears to hear. He insisted on having the truth in the form of an official schedule.

On the 27th, the day on which von Lersner communicated his melancholy report from Headquarters, von Hintze expressed to Hertling his fear that a catastrophe might be impending. The old gentleman, suspecting nothing, called Hintze a pessimist. Hintze himself, however, "did not consider that his view, although gloomy, warranted any new steps, since it had not yet been confirmed by General Headquarters". On the following day von Hintze went to Spa in order to ask for "an unequivocal and definite declaration on the military position". With this declaration in his pocket he would at last be ready to "suggest any necessary action". The last hour of the Empire had come; the alliance of the Central Powers was about to collapse; General Headquarters were loudly appealing for help; and the director of the foreign policy of Germany had no other care than to act in accordance with bureaucratic precedent. But without an unequivocal and definite declaration from the Generals no bureaucracy could come to a decision. The final chapter of a great historical drama was administered in accordance with official routine.

It was on the 28th of September 1918 that Ludendorff saw himself at the end of his resources. At six on that evening he left his office, went down to Hindenburg on the floor below, and there informed the Field-Marshal that the collapse of Bulgaria had destroyed his last hope. Even if the Western front remained unshaken the position could only deteriorate, and the one duty of Headquarters would be to act firmly and rapidly in insisting upon an armistice being offered to the enemy. Deeply moved, Hindenburg told Ludendorff that he himself had come to the same conclusion. Both Generals had the idea that the armistice would follow historical precedent and would allow them to evacuate the conquered territory in good order, halting once the German frontier was reached. They took it for granted that the East would remain untouched on the assumption that the Allies would be afraid of Bolshevism.

Ludendorff had not reached his conclusion on the spur of the moment. His belief in victory until the 8th of August might be compared to an immovable front; on that date this front was broken. Reviewing these events, we may ask why, during the council on the 14th of August, Ludendorff insisted on the advisability of holding the line in the hopes that this was the way to ultimate victory, whether he believed in his own assertions, and what was his general view of the future thereafter. How did he, as an expert, view the prospect? In any case even the moderate degree of confidence which Ludendorff displayed on that occasion was misunderstood in his surroundings. As a matter of fact, Germany's position was such that only a miracle could save her.

Who was to be the author of the miracle: the German soldier or the German diplomat? As an expert, Ludendorff knew that the German troops had come to the limit of their endurance; it was his mistake that he believed in a diplomatic miracle. He honestly believed that Herr von Hintze would shortly succeed, with the help of a neutral Power, in inducing the enemy to enter into negotiations, and on this assumption he had every right to believe that the front line would hold. Both his confidence and the strategy which was dictated by it were based upon this belief.

If negotiations were entered upon without delay he was justified in this strategy, with its assumption that the German line would not be forced to yield. As a soldier, Ludendorff had a poor opinion of diplomats; but in the hour of need he expected a miracle. The miracle did not occur; and meanwhile the military position deteriorated until a catastrophe was to be expected any day. A final break-through and collapse was prevented only by the amazing steadfastness of the German troops; and in the end it was the military expert, the officer with his cool and unemotional methods of thought, who appreciated the hopelessness of continuing and determined to ask for an armistice.

Dr. Rosenberg's perfectly unbiased and objective investigation comes to the conclusion that it would have been much more in the personal interest of Ludendorff not to insist on asking for an armistice, but to fight to a standstill in the rôle of the "invincible hero". "Once the catastrophe had arrived it would have been easy to find a scapegoat, and to carry on the heroic legend. In 1814 Napoleon, in a somewhat similar position, fought to his soldiers' last drop of blood. But although Ludendorff fully appreciated his own merit, he had no egoistic feelings of personal domination; and it was only the horrible confusion of German politics which compelled him to play dictator for three years. In September 1918 he did not think of his personal future, position, or historical reputation, but only of the welfare of his troops, as a good colonel will think of his regiment first, and of himself in the second place."

On Sunday the 29th of September, at ten in the morning,

Hindenburg and Ludendorff met Herr von Hintze at the Hôtel Britannique at Spa; beside them only Colonel Heye of General Headquarters was present. Hintze described the threatening position in Germany and gave an account of his efforts at The Hague. He was followed by Ludendorff, who gave an exposition of the military position, and ended by claiming that an immediate armistice was essential. Herr von Hintze was alarmed, and said that something must be done in order to alleviate the dangerous results of such a rapid transition from paeans of victory to dirges of defeat. What Hintze had in mind was a dictatorship or a revolution from above, an idea to which Ludendorff objected on the grounds that there can be no dictatorship without victory. Hintze further suggested an offer of peace to President Wilson, and to this suggestion Ludendorff agreed. When Hindenburg went on to say that the terms of peace must include the annexation of Briey and Longwy, Ludendorff cut him short by saying that the time had passed for that.

On a later occasion Herr von Hintze thought it necessary to offer a defence for his conduct in accepting and complying with the desires of the Generals for an immediate armistice. As far as the later events are concerned, all that matters is the fact and not the motives. While the Generals were conferring with von Hintze at the Hôtel Britannique, the Kaiser arrived, having travelled from Herbesthal to Spa in the company of Lt.-Col. Niemann, and unaware of Ludendorff's declaration. At half-past eleven the Generals and Herr von Hintze waited upon him, and the Secretary of State made a report upon the external position. The Kaiser, however, wished to hear about the position in Germany. This was a subject on which von Hintze did not feel competent to speak, and he did not give a report on the internal position until the Kaiser had requested him in unmistakable terms. Finally Hindenburg described the position of the army, and concluded with the admission that an immediate armistice was essential. Ludendorff briefly confirmed him.

The Kaiser preserved a calm appearance and asked for suggestions. von Hintze thereupon began to speak about a dictatorship, whereupon the Kaiser said: "A dictatorship is nonsense". The Foreign Minister now suggested that it might be possible to "canalise" the threatening revolution, and offered to suggest a peace conference to President Wilson, coupled with a request for an immediate armistice as demanded by the military leaders. To this suggestion the Kaiser agreed, but when Hintze went on to ask that he might be allowed to tender his resignation, since his reactionary reputation would make his position impossible in the Government, the Kaiser declined to accept the suggestion. It was Ludendorff who had emphasised the necessity of immediately offering an armistice, a demand reached after mature reflection, and backed by Hindenburg. The Kaiser and the Foreign Secretary, who represented the Chancellor, agreed with the Generals. Thus here again the triumvirate was unanimous.

At noon the Chancellor arrived. On being informed of what had passed he was "profoundly shocked" and remarked to his son, "This is terrible". In the afternoon he tendered his resignation. His successor carried on the tradition of political naïveté initiated by him.

In the afternoon the Kaiser said to von Hintze in conversation: "According to Hertling there is not such a great danger of revolution after all. There is no hurry about peace or a new Government; we must think things over a little." Hintze, however, supported Ludendorff's standpoint, which was the only possible one in the circumstances. His idea was that the Allies would refuse to negotiate with the old rulers of Germany, and that an armistice was impossible without a definite change in the German constitution.

The Chancellor's office had already prepared a proclamation announcing a popular Government, together with Hertling's resignation. This proclamation was lying on the table. Hintze did his best to persuade the Kaiser that its promulgation was an unavoidable step. The Kaiser hesitated and made for the door; von Hintze followed him and reiterated his request. The Kaiser finally turned back, went up to the table, and signed the proclamation.

Hintze was insistent because he felt himself bound to Ludendorff. He had informed the General that the new Government would be formed by the 1st of October, and the offer of an armistice signed by the new Government was to be despatched on the same day. At 9.40 P.M. Hintze issued instructions from his office at Spa to inform Vienna and Constantinople confidentially that Germany intended to ask President Wilson to mediate, and to ask for an immediate armistice. During the night Hintze, accompanied by Count Roedern and Major von dem Bussche, proceeded to Berlin. The Major had been instructed by Ludendorff "quite frankly to inform members of the Reichstag and Government departments of the gravity of the military position". Rapid decisions would be necessary, and it was desirable that they should have some facts upon which to go.

On the evening of this eventful day, at 10 P.M., Ludendorff met the officers of his staff. He spoke clearly and quietly, pointing out that since the collapse of Bulgaria all hopes had vanished of holding the Western front by drawing divisions from the East. "We have no reserves to transfer to the West. . . . I should be not better than a gambler if, in view of the gravity of the position, I did not insist upon ending the war by asking for an immediate armistice. This has been done. I have come to this conclusion in complete agreement with the General Field-Marshal. A Ministry on a broad basis is essential to conclude the war. The Field-Marshal and I have the greatest reluctance in making this statement, but it is essential that it should be made. . . ."

During the subsequent events Ludendorff did his utmost to assist in the formation of a popular Government. While Hertling saw something monstrous in the effort to introduce parliamentary Government into the Empire of the Hohenzollerns, Ludendorff saw that the change was necessary, but also perceived that the constitutional revolution could only be brought about by order. The foundations of the new Germany were not laid by the Reichstag, still less were they won in conflict; Ludendorff gave instruction for them to be laid.

Dr. Rosenberg rightly says that this revolution is unique in history. "There are examples of military rulers and dictators voluntarily giving up power; but history has never before witnessed a dictator doing his utmost to put his enemies in power, in the way in which Ludendorff did at the end of September and the beginning of October 1918." Nevertheless this quaint phenomenon is intelligible. The nation was accustomed to receive orders and to obey, and it was incapable even of effecting a revolution otherwise than by order. Apart from Ludendorff there was nobody at this moment who had the necessary initiative and energy to give the necessary orders. Hintze, too, complied with instructions, and eagerly strove to call into existence a new popular Government. He had returned to Berlin on the morning of the 30th of September and immediately called upon the Vice-Chancellor, Herr von Payer, whose obvious task it was to form a new Government. von Payer received the news about the events at Spa as "a surprise of the most terrible nature".

His recollections of these events allow us to see the effects of the Spa revolution in the inner circle of the Cabinet. The utter disillusionment, which was equally pronounced in the Reichstag, constitutes the ground of one of the chief charges against General Headquarters, who are continually accused of not having given anything approaching an accurate version of the military position to the Government or the people. Erzberger, like Payer, spoke of a terrible surprise, Stresemann of a brusque and unexpected turn, and von Heydebrand exclaimed in the corridors of the Reichstag: "We have been betrayed".

This universal surprise manifests a peculiarly German quality. With a few insignificant exceptions there is no general in history who in a position similar to Ludendorff did not retain some glimmer of hope or at any rate pretended to some glimmer of hope. It is definitely established that the Vice-Chancellor and members of the Reichstag were free at any moment to go to Headquarters or to the front, and to gather more than merely superficial information and form an independent judgment. There were, in fact, instances of politicians appearing at Headquarters, and some of them actually ventured to the front. But the memoranda made by these guests about their expeditions uniformly show that they did not venture to form an independent opinion. Here again we meet with the original cause of the German tragedy: the enormous prestige enjoyed by the military experts compelled even those politicians who were capable of independent judgment to preserve a deferential attitude. Thus the Crown Prince's memoirs show that von Heydebrand was correctly informed during his visit in July 1918, and there were other instances where a man of average intelligence could have learnt what was the actual position. It had become a matter of orthodox belief that the Western front would and must hold. Apart from the official reports, nobody attempted to investigate facts for himself, still less was there anybody possessing sufficient constructive imagination to draw the right conclusion from the simple fact that the military position had changed in the summer of 1918. If the decision taken at Spa on that Sunday came on the politicians like a bolt from the blue it was certainly not the sole fault of General Headquarters.

Meanwhile Ludendorff had used the telephone to keep von Payer on the move, and on the 30th of September the latter concluded his mission by suggesting Prince Max of Baden for the post of Chancellor. "It is characteristic of the vacillation and uncertainty of the majority parties", Rosenberg remarks, "that they did not possess among their numbers a single man fit to lead the Government. The period of democracy in Germany begins with a Prince from the south of Germany taking the position of Chancellor."

And indeed there is no historical justification for expecting anything better of the Reichstag. We saw how the Reichstag proved a failure at decisive moments of the war and avoided every proffered opportunity of seizing power; it was only to be expected of this pitiable, undecided, and hesitant body that its last gesture towards the old rulers should be one of obeisance. There is a kind of hypothetical history which delights in assuming the possibility of the impossible. Its authors have not spared the inauguration of the new Germany, and have outlined a picture full of heroic deeds where gallant leaders of the populace seize power and call upon the people for a last act of resistance, statesmen and improvisers of genius grow on the soil of popular commotion, and the war eventually is carried on with renewed forces. But such ideas are dreams. This kind of history has to a certain extent blinded people to facts and given rise to pernicious legends. We said above that every fighting nation bears its own history "under its helmets"; and where a whole nation is wont and delights to obey, these characteristics will not be lacking at a critical moment. A defensive action on a grand scale was possible in Germany only if prepared beforehand, organised, and initiated by the General Staff; and we have seen the reason why such a defensive action was not undertaken. An improvised *levée en masse* is impossible in Germany.

Ludendorff had fixed the 1st of October as the last day for making an offer of armistice. Finding the appointment of Prince Max delayed, Ludendorff consented to extend the time limit to the 2nd, under the proviso that the new Government should be formed by then. Prince Max arrived at Berlin as Chancellor-Designate on the afternoon of the 1st. He held a completely different view of the position, and believed that he had complete freedom of political action; accordingly he was surprised and alarmed on learning that the request for an armistice had already been determined upon and that, in fact, the request was to be despatched to the foreign Governments by the morning of the following day at the latest.

There could be no doubt as to the gravity with which the position was viewed at General Headquarters. During the forenoon of the 1st, two telegrams from Headquarters had arrived at the Foreign Office, the first of which, despatched by von Lersner, contained an urgent request on the part of Ludendorff to despatch the armistice offer without delay ("I can hold the troops to-day, but I cannot foretell what will happen to-morrow"), while the other, despatched by Freiherr von Grünau, contained the same request, and went on to say that the offer was to be despatched in any case, even if there was a delay in forming a Government. "A break-through may take place at any moment, and Ludendorff says that he is feeling like a gambler."

Despite the urgent language of these telegrams the Prince felt reluctant to execute the resolve come to by the Crown Council at Spa. In his opinion the immediate offer of an armistice was not the correct method of initiating peace negotiations. His dilemma was indeed a difficult one. The anxiety at Headquarters to take precautionary measures against the catastrophe was so great that Ludendorff was unwilling to wait until a new Government should be formed, while on the other hand the Kaiser sent a telegram vetoing the suggestion that the offer might be signed by the old Government if it was impossible to form a new one.

As yet Prince Max had not received the Grand Duke of Baden's sanction, without which he was unwilling to accept the office of Chancellor. Accordingly Major von dem Bussche, the representative of General Headquarters, stopped the train in which the Kaiser was proceeding from Spa to Berlin when Cologne was reached. The train had a telegraph coach by means of which connection could be established with Berlin and Karlsruhe, so that it was possible simultaneously to obtain the sanction of the Kaiser and the Grand Duke. At midnight Prince Max was Chancellor at last.

Before this step had been taken Ludendorff had sent yet another telegram to Berlin, enquiring about the telegraphic connections between Berlin, Berne, and Washington, and urging the necessity of taking all possible steps to avoid waste of time. "From Berne the offer must be sent to Washington without a moment's delay. The Army cannot wait another forty-eight hours. . . . I urgently request that nothing be left undone to have this offer transmitted as quickly as possible. It is of the utmost importance for the offer to be in the hands of the Allies by Wednesday evening or Thursday morning (3rd of October)."

The 2nd of October began with a meeting of the party leaders. At the request of von Payer there were present Stresemann (National Liberal), Gröber (Centre), Count Westarp (Conservative), Ebert (Social Democrat), Fischbeck (Progressive), Gamp (Reichspartei), Haase (Independent Socialist), and von Seyda (representing the Polish interest), to whom Major von dem Bussche proceeded to give an exposition of the military situation. The Major strictly obeyed Ludendorff's injunction to confine himself to essential facts. Beginning with the fact that the collapse of Bulgaria had upset all the calculations of General Headquarters, he went on to the conclusion that it had become necessary to take up the defensive in the West, but pointed out that the engagements of the last six days had been carried to a satisfactory conclusion despite losses of men and material. "The great bulk of the troops have behaved extremely well; their efforts have been almost superhuman. . . . Nevertheless, Headquarters has been compelled reluctantly to declare that in all human probability there is no hope of compelling the enemy to accept our terms. The decisive factors consist in the enemy tanks and in our lack of reserves. . . . At the moment the German Army is strong enough to resist the enemy for months . . . but with every day the enemy comes nearer to his goal, and becomes proportionally less inclined to conclude an acceptable peace. For this reason we cannot afford to lose time." The Major concluded with the following words: "While the offer of peace is made, we must present a united home front, making it clear to the enemy that we are firmly resolved to continue fighting in the event that he is unwilling to offer us terms of peace, or other than humiliating terms. In such an event the conduct of the Army will depend upon the firmness shown at home and on the backing it receives on the home front."

The Major was simply expressing the views held at Headquarters. But while Ludendorff was free to decline a humiliating peace, he could not call up the nation for a last despairing rally; even the blindest obedience cannot bring about a miracle. In themselves the events are simple. The Generalissimo, representative of the Prussian military academy and the will to win, informs the Reichstag that the war is lost, and leaves it to the representatives of the people to determine the further fate of the nation.

Colonel Schwertfeger, whose official description of these events is contained in the report published by the Reichstag Committee of Investigation, describes the party leaders as being "completely unnerved" by this exposition. von Payer's memoirs refer to the "consternation" of the audience. "Count Westarp and Stresemann declined to believe that our position was really so desperate. . . . What caused the panic among the party leaders was not any particular detail in the exposition, nor the manner in which it was delivered, but the fact that the hopelessness of our position was shown up with brutal plainness by the facts communicated. . . . Up to this point the parliamentarians had formed their judgment from the daily reports and they were absolutely unprepared for the blow. . . . After a brief discussion . . . they separated in great consternation." There is no need to adorn this tale.

Meanwhile the Kaiser and Hindenburg had arrived at Berlin, while Ludendorff remained at Headquarters. In the general course of the German disaster this point is worth noting. Ludendorff, the dictator against his will, had surrendered his power to the popular representatives. But the meaning of this action had not been understood. It appears that an acquired talent is required in order to understand historic actions, and the German parliamentarians were without this talent. A great event had taken place; but neither Ludendorff's admission of defeat nor his appeal to the nation found any echo in the shape of political action; their only result was a depression which was of a purely personal and physiological nature. When the great moment came the politicians became so many private individuals, who reacted by displaying private distress instead of political activity.

The fact is that a political will requires certain conditions before it can exist. The history of Germany had not provided these conditions; a German Gambetta is the product of a literary imagination. von Payer, while executing the popular will of Germany, never suspected that Germany itself had changed.

The decisive consultation took place at six in the afternoon at the Chancellor's official residence under the Chairmanship of the Kaiser. The new Chancellor, Prince Max, requested in urgent terms that no armistice should be offered as yet. The Kaiser declined, and preferred to abide by the resolution reached at Spa; nor can he be blamed for this decision unless it is held that a *levée en masse* was possible—an idea which is a romantic fiction and having nothing to do with history. With the exception of the Prince and Dr. Solf, the Government was unanimous in deciding that the only course was to offer an armistice to the enemy.

Prince Max, however, was not satisfied; the idea of being legally and historically responsible for such a step terrified him. His objections were summed up in a *note verbale* addressed to the Field-Marshal. In his opening speech as Premier he intended to outline a peace programme "in close but not slavish imitation" of Wilson's fourteen points, and to invite all the Powers to a peace congress.

Such a plan, however, required two days to be carried into practice, and it was doubtful whether Headquarters could grant this time. In such an event Prince Max intended to despatch a note to all the enemy Powers without any accompanying speech, in which case, however, he required General Headquarters to state in writing that the position on the Western front was such as to make it impossible to postpone the offer as late as Saturday the 5th of October.

Hindenburg replied to the Prince's note on the 3rd in a letter in which he said that Headquarters must insist upon the immediate despatch of the peace offer. "The position is growing daily more desperate, and it may be necessary to make the most far-reaching decisions at any moment. It is essential to break off hostilities and to avoid useless sacrifices. Every day that is lost means the death of thousands of our brave soldiers."

Together with the *note verbale* the Prince had submitted five questions to Hindenburg with a request for answer. The reply to these was given by Hindenburg after telephonic conversation with Ludendorff, at a meeting of the Council of Ministers held on the evening of the same day. Substantially his remarks agreed with the contents of his letter. One of the questions with which he had to deal asked whether General Headquarters realised that peace negotiations, entered upon in a desperate military position, might lead to the loss of German colonies and of German territory (Alsace-Lorraine). The following was Hindenburg's reply: "Headquarters are willing to envisage the surrender of small parts of Alsace-Lorraine with a French-speaking population. The surrender of German territory in the East is out of the question." If the enemy demanded more Germany must continue to fight to the last man. Count Roedern, a brilliant sceptic, objected that while it was possible for a single battalion, a forlorn hope, to fight to the last man, this was out of the question for a nation of 65 millions. Headquarters, however, insisted upon their demands undeterred by the prospect of being compelled to continue hostilities.

The five questions outlined by the Prince revealed a gloomy prospect without making the next step any clearer. It still remained doubtful whether the armistice request was to be despatched immediately, or whether the Chancellor was to be given two days' grace. Hindenburg began to feel doubtful.

It was von Hintze who eventually brought about a decision by drawing attention to the position at the front and to Ludendorff's urgent appeal. The Prince now gave up thoughts of resistance; but, while no longer insisting upon a delay, he still urged the desirability of despatching an offer of peace rather than a request for an armistice. Colonel von Haeften thereupon telephoned to Ludendorff, who replied that he must still insist upon the offer of an armistice being made and being despatched without delay. "At the moment the position is not threatening, but at any time a general attack on the whole front, in the West, in Italy, and in the Balkans, is to be expected. The German Army is urgently in need of a rest, and once such an attack is delivered it may be of decisive importance whether this respite is given twenty-four hours earlier or later."

Ludendorff's reply is plainly based on a false analogy drawn from the example of the wars of an earlier and more romantic age, when an armistice was not necessarily the preliminary to the termination of hostilities. In 1918, however, analogies drawn from past wars were no longer applicable.

Ludendorff's reply brought about an agreement between Headquarters and the new Government. Within a few hours the German offer of peace and request for an armistice would be known throughout the world. The actual note was drawn up in the afternoon of the 3rd of October, and the only alteration which Ludendorff wished to introduce was a verbal one intended to emphasise the fact that Wilson's fourteen points were "to serve as basis of the negotiations". In its final form the tenor of the note was to this effect, although the word "serve" was omitted.

The note was despatched to President Wilson in the small hours of the 4th. In this document the German Government requested the President of the United States "to work for peace" and to invite all the belligerent states to take part in the conference. "The German Government accepts the programme drawn up by the President as basis of the peace negotiations. In order to avoid further bloodshed the German Government requests him to bring about the immediate conclusion of an armistice."

Events now began to take their irrevocable course. On the following day Prince Max informed the Reichstag of the Government's step, concluding his speech by expressing his confidence that the German people would reject intolerable conditions and would, in such a case, "continue to fight to the last man. I feel no tremor at the thought of such a contingency. I know the enormous latent resources which even now exist in the nation."

In his reply President Wilson required the evacuation of occupied territory, while leaving open the question of negotiations. Hereupon Prince Max requested Ludendorff to come to Berlin, as he wished to obtain a clear idea of the fighting powers of the combatant forces. The Prince required to know whether there was an immediate danger of a break-through, and whether there was any possibility of the German front being rolled up and the German Army annihilated in a pitched battle. The Prince's anxiety to have an answer to these questions is intelligible and justifiable.

The position had not changed, and was as hopeless as it had been ten days ago; Ludendorff could only repeat his previous remarks. The Prince thereupon expressed a desire to hear the opinion of the Army Commanders and their Chiefs of Staff; Ludendorff, however, objected to this appeal to other officers than himself, and pointed out that outside General Headquarters nobody possessed a comprehensive view of things. It would be unwise to draw conclusions upon the general position based on the experiences of any one army.

The same questions were discussed at a meeting of the War Cabinet held on the 5th of October. On the same day Walther Rathenau published an article in the *Vossische Zeitung* demanding a *levée en masse*. In the Cabinet as then constituted, Dr. Solf was Foreign Minister ; Herr von Payer, Vice-Chancellor; Herr Trimborn, Home Secretary; other members of the Cabinet were Herr Bauer (Social Democrat); and Haussmann, Erzberger, Gröber, and Scheidemann. The Cabinet was busily engaged in conversations with Wilson and was not in a position to pass a resolution in favour of a *levée en masse*; on the 12th of October the second German note was despatched to America.

Meanwhile the battle which had begun at the end of September continued. It was Marshal Foch's ambition to break through the German front, his aim to force a strategic victory. From the 17th to the 20th of October, Crown Prince Rupprecht's Army Group retired to the Hermann line, which had not been properly completed as yet; simultaneously the Crown Prince's Army had retired to the Hunding-Brunhild line. Ludendorff now began to urge the necessity of beginning the construction of the Antwerp-Meuse line. Unfortunately all available forces were at work on the Hermann line, and the lack of prepared lines in the rear, coupled with the necessity of fighting defensive actions in practically unprepared lines, entailed great sacrifices upon the troops. Thus in the beginning of October General Boehn's Army Group reported to Headquarters: "During to-day's engagements we had to draw upon our last reserves. The entire front is being held by depleted and exhausted units. A moment will inevitably come when an enemy attack will develop into a break-through. . . . An immediate retirement to the Antwerp-Meuse line is an imperative necessity."

On the 17th of October Ludendorff was requested to attend a Cabinet Council. Wilson's reply to the second German note had arrived, in which the President demanded the cessation of submarine warfare together with constitutional reforms within Germany. The Chancellor turned to Ludendorff and explained that before replying to the note the Government must be informed of the state of the military defence of Germany. Ludendorff replied: "War is full of probabilities and improbabilities. Nobody knows what will happen ultimately. . . . The fortune of war plays a part of every campaign. Perhaps fortune will favour us again."

As an outline of policy this was somewhat inadequate. The Government was particularly anxious to know whether any possibility remained of drawing forces from the East. In Russia and Roumania only 26 divisions remained, consisting mainly of comparatively elderly men, and the debate turned towards the question whether an occupation in the East was necessary, and whether the Ukraine was important as a source of supplies. Scheidemann eventually exclaimed: "Better an end with terror than terrors without end", while Conrad Haussmann advocated an appeal to the nation.

Ludendorff made a second attempt to explain himself. "A break-through is possible but not probable; on the other hand, a change for the worse can take place at any moment. The front is neither more nor less firm than it has been; the aggressiveness of the enemy appears to be on the wane. The negotiations with Wilson have not yet led to any result, and we are free to take what steps we like. We are honestly trying to reach an agreement; but surely it is no crime to give up the attempt if the enemy demands the impossible. If it comes to extremes—if we are asked to surrender at discretion—it will become our duty to continue hostilities. The German nation will lend its last forces to the Army." Ludendorff further opposed a suspension of the submarine campaign, and urged upon the Chancellor the expediency of placing Herr Ebert in a conspicuous office in order to win the confidence of Labour by conferring office on a Social Democrat.

This speech constitutes Ludendorff's so-called "breaking point", politically and intellectually, as Dr. Solf did not fail to remark in the same session. Yet Ludendorff had merely been saying what everyone had been thinking. A soldier, a military expert, and a strategist, he had no illusions, and was perfectly aware that the position of Germany could not improve. Disaster was impending in the South-East, and the Austrian surrender was only a question of hours. All these matters were no longer subject to doubt.

The Chancellor's questions were searching and painful, but only too intelligible in view of his mental strain. Ludendorff answered them truthfully. Any day a disaster might occur and the amazing power of the German infantry collapse; on the other hand, the war might drag on for months. Such were the facts which, on the 28th of September, had compelled Ludendorff to ask for an armistice, and these facts had not changed. The war was lost; but if an attempt were made to impose humiliating or crushing terms upon the nation, a last and desperate fight would become necessary in spite of all the rules of strategy.

Delbrück has described Ludendorff's attitude as ambiguous. Yet the fact is that far from being blameworthy his position was perfectly simple and dictated by a characteristic crudeness and lack of psychological insight. It might be condemned for reasons of policy, as in fact was done by the Government, but it is not deserving of moral censure. After his own fashion Ludendorff was consistent and true to the Germany of which he was a child. He believed in the figure of the heroic Teuton as handed down by the Church and by Jingo historians: he had absorbed this brave heroic panorama, it had become his world; and in appealing to the German nation for a heroic gesture he was merely testing the genuineness of his vision. In a similar juncture Luther had opposed his iron obstinacy to the world; and Ludendorff's ideas, in spite of a certain military rigidity, were not without historical and psychological elements. Instinctively rather than as a matter of rational reflection he felt the difficulties standing in the way of a German *levée en masse*. A people grown up in obedience and accustomed to receive commands was incapable of any elementary reaction; alone among the nations, Germany had not emerged beyond the *status pupillaris*. It is for this reason that Ludendorff had to give orders for a revolution; the popular leaders were incapable of attaining power except by command. But beyond this, Ludendorff was powerless; from now onwards, so he held, it was the turn of the popular leaders to give orders.

Military critics have spent much labour on the question how long the German Army could have resisted and whether a continued defence would have led to that pause which was essential if the Army was once more to regain steadiness. In his report General von Kuhl expresses the opinion that the *morale* of the German Army in 1918 was such as to have rendered a continued resistance possible: "after deducting deserters and shirkers there remained a healthy core which might have offered vigorous resistance". In Kuhl's opinion the Army would have had time in November to settle in the Antwerp-Meuse line, with the possibility of continuing its defence further in the rear.

The fact is that the Allies had lost breath; the roads and railways having been destroyed, the advance came to an end; the American reinforcements broke down altogether, and some time had to elapse before the rationing system of the British Army worked smoothly again. The position was similar with the French and the Belgians. As the German line was threatened their power of resistance grew, and Field-Marshal Haig admits that during those weeks the German Army had retired, defending itself vigorously and in good order.

While all this is true, it does not touch the core of the problem. The only question to which it suggests an answer is whether a prolonged resistance might have led to more satisfactory terms of peace. That this might have been the case does not appear probable from the writings of enemy statesmen and generals. The Allied armies would have resumed the attack as soon as the difficulties of transport and rationing were overcome; the American military machine, which had only just begun to function properly, would presently have worked at full pressure. Thus it was hardly to be expected that the Allied forces would lose the will to fight at the precise moment when they reached the German frontiers. A continuation of the war would have transferred the scenes of battle into the Rhineland, the Tyrol, Bavaria and Saxony.

In these matters it was safe to rely on Ludendorff's military judgment; while it was possible to prolong the war, it was not possible to improve the position of Germany, and in fact General von Kuhl describes the possibility of a prolongation of the war only in order to be able to say that the revolution "shattered the sword in the General's hand". But such an assertion serves no good cause, and least of all that of Ludendorff. The latter never entertained the absurd notion of prolonging the resistance for months with the exhausted and weary forces of the armies in the field; both this idea and the fiction of the "stab in the back" of the field army were created at a later date.

The truth is that Ludendorff, alarmed at the prospect of harsh terms of peace, did demand a continued resistance, but in so doing he had in mind a *levée en masse* bringing further reinforcements to the existing army. Even so, however, he knew that the war could not be won; at the same time "a popular rising could not but improve our position". It was for this reason that at the Cabinet Meeting of the 17th of October he appealed to the popular leaders with the words: "You must seize hold of the populace and carry them with you". Such a demand was inspired by ignorance of the German character and the historical position; but it is also completely different from the theories evolved by General von Kuhl after the event, when he imagines that a continuation of the war would have been possible without a popular rising and without any reinforcements for the exhausted army.

After the Cabinet Meeting of the 17th of October, Ludendorff returned to Spa. Previously Prince Max had thought the German method of opening negotiations for peace mistaken; now, however, he agreed with Ludendorff that the only answer to a demand for surrender at discretion would be a *levée en masse* on the part of the entire nation. On the 22nd of October he said: "The very fact that we are willing to accept an equitable peace implies a duty not to accept a dictated peace without a struggle. A Government failing to appreciate this distinction would deserve the contempt of the fighters and workers of the nation."

On the following day Wilson's Third Note was received. It revealed the cruel fact that an armistice would only be granted at the price of unconditional surrender. Accompanied by Hindenburg, Ludendorff immediately hastened to Berlin.

On the afternoon of the 15th of October the two leaders had an audience of the Kaiser. Once more Ludendorff sought to sum up his reasons for continuing the conflict: "If there is a popular rising, the war can be continued for some months. A fortress that surrenders without having defended itself to the last is dishonoured." Following constitutional practice, the Kaiser referred the Commander-in-Chief to the Chancellor. Prince Max was unwell, and Herr von Payer, the Vice-Chancellor, received the Generals in the presence of Admiral Scheer and General Scheuch, the Minister for War. There followed an excited discussion. Ludendorff pointed out that it was in vain to repose any hopes in Wilson and described the terrors of impending revolution. The last moment had come; if the Government accepted President Wilson's note without calling upon the nation for a last resistance, the fate of Germany would lie in the hands of the President. The Government was no longer free in its decisions, and the private opinion of Prince Max as to the necessity or possibility of a national rising no longer carried weight. The decision had been reached when Herr Erzberger had conveyed the decision reached by the Centre Party, which declared unanimously against a national rising. Conrad Haussmann was the only member of the Government to share the Chancellor's views; all the others, Dr. Solf, Trimborn, Erzberger, Scheidemann, Bauer, and Gröber had given up all thoughts of a national resistance.

On the same day (25th of October) on which Ludendorff attempted to impose his views on the Government, General Headquarters had issued an appeal to the Army. The troops were informed that Wilson's note demanded the submission of Germany and claimed the right to control the constitution of the country. "Wilson's reply is such that no soldier can accept it. It proves that our enemies spoke of an equitable peace only in order to deceive us and to break our will to resist. The only possible reply can be an appeal to continue our resistance to the last ounce of our forces."

This appeal had been signed by Hindenburg and sanctioned by Ludendorff in the belief that it had the Government's cognisance and approval. Such, however, was not the case; and although Colonel Heye gave orders for the appeal to be stopped, he could not prevent the Reichstag from interesting itself in the question. The appeal had been mentioned during the Press Conference; it had been overheard at the telephone office at Kovno, and had been communicated to the Independent Socialist Party. In this way the Reichstag heard of the appeal; intense indignation was felt, and the Army Order was criticised as a piece of unjustifiable breach of privilege.

On the evening following his interview with the Vice-Chancellor, Ludendorff was informed that the wrath of the Reichstag was turning against him. He thereupon wrote a letter of resignation; it was now 8 o'clock in the morning of the 26th of October. He referred to the news just received that the Government had declined to entertain the idea of a *levée en masse*, pointed out that he himself was looked upon as a supporter of the war, and suggested that his resignation would ease the position for Germany, and concluded by begging leave to tender his resignation. He had not finished when he was joined by Hindenburg, who saw the letter of resignation and succeeded in persuading Ludendorff to change his mind. The two Generals thereupon tried to call upon the Chancellor, but were informed that he was unwell and not at home to anybody.

While Ludendorff was waiting for the Prince's reply, urgent news was brought by Colonel von Haeften to the effect that the Government had prevailed upon the Kaiser to dismiss him. Within a few minutes Hindenburg and Ludendorff were asked to attend at Bellevue Castle, where they found the Kaiser's manner completely altered. Pointedly addressing Ludendorff, he criticised the Army Order. Ludendorff replied that it was his painful impression that he no longer enjoyed the Kaiser's full confidence, and asked leave to tender his resignation. The Kaiser nodded; Ludendorff was dismissed.

The above scene had only lasted a few minutes. Ludendorff quitted Potsdam alone. In the offices of the General Staff he took leave briefly from Hindenburg; that same evening he proceeded to Spa by train. There followed a painful farewell from the officers of his staff and from the scene of his labours. On the following day, the 28th, he reached Aix-la-Chapelle, his first Headquarters during the war. Memory carried him back to Liége.

CHAPTER XII

THE END

THE tragedy of Ludendorff, the Prussian General who had been dictator of Germany for two years, was followed by an epilogue. The collapse of the Empire found him at Berlin; when the revolution came he escaped, "with a false beard and a pair of dark glasses, in the dark", to Denmark, whence he went to Sweden, where friends received him at their estate, Hessleholm. The manner in which a man tries to protect himself against the bullets of political assassins is of no essential importance, and if this were all, the false beard and the dark glasses would in no wise discredit Ludendorff; Napoleon, in a similar position, disguised himself under an Austrian hat and a Russian military cloak. Reviewing these days, Ludendorff said, on one occasion, that it was "the silliest thing the revolutionaries could have done not to have despatched him." This saying has nothing in common with Napoleon's dying words. When Bonaparte said to Bertrand that he ought to have died at Borodino he was actuated by considerations of posthumous fame.

Such regrets were alien to Ludendorff; they were, indeed, impossible, for they would have required a critical self-analysis of which Ludendorff was incapable. His feeling of inner certainty, his general opinions and judgment, survived the collapse of Germany. The German effort, the unparalleled resistance which she had opposed to the utmost efforts of the greatest world Powers for fifty months, and the recollection of the victories which were and which might have been achieved, gave rise in his mind to a cloudy and heroic picture which was impervious to criticism. Again, he had no sympathy with any attempts to interpret the collapse of

Germany as the work of Fate, or to find solace in the belief in a blind and tragic destiny. He was shaken by no shadow of self-reproach or of doubt, so that it could not but be that he would continue on his way, which would lead him to wage war against those who had spoilt his war. He had complete faith in his mission and its success.

Thus it came about that he emerged again during the Kapp Putsch. On the 8th of November 1923, he was to be seen at the Bürgerbräu at Munich; on the next day he marched at the head of the force which, under Hitler, was hoping to capture Berlin, to depose the Government, and to set up a dictatorship. Brought before the Courts he confessed himself a Monarchist, but also declared that there could be no monarchy without a previous popular dictatorship. He reminded the Court that he had been blamed for not having made himself dictator during the war; and he admitted that had he done so Germany would have been saved much trouble. He was eventually dismissed alone among the accused. He replied by calling the verdict "an insult to his uniform and decorations". Having made this utterance he returned to his desk.

The following years revealed much that was unsuspected in Ludendorff, a spectacle which cannot be explained by the comforting theory that his behaviour was natural in a man who had fallen from a great position and was no longer capable of rational thought, or in contact with the world. There is much in Ludendorff's subsequent career that may be explained as the aberration of a powerful but impotent will or of wounded *amour propre*; but at the same time ideas and intentions become apparent in the gloomy landscape of his soul, which formed a fundamental part in the volitional make-up not only of Ludendorff, but of a large class of the German nation, and whose force, far from being exhausted, is more active beneath the surface to-day than ever before.

Reduced to a single formula, Ludendorff's idea of the world, and of existence in general, may be described as the belief that life is strife. In each nation there is conflict between classes, between interests, and between parties, with power for goal; between the nations there is conflict for hegemony. Plainly this is a part of the nature of things and a divine dispensation. Education and higher ethics may apply a kind of code to the struggle for power, while science may discover the most suitable form of war; in any case it will never be possible to abolish force, since such an abolition would be contrary to human nature and to nature in general. Even the good can only maintain itself by force.

There is nothing novel and much that is dreary in accepting empirical facts and deriving thence the entire sense and meaning of the universe; but Ludendorff is more consistent than literary worshippers of force, and indeed, as Aldanoff remarks, "differs favourably from Professors of International Law and European Press Magnates" in avoiding any attempt to represent his enemies as monsters of wickedness and treachery. Where the enemy displays a more powerful will to destruction, and a more ruthless manner of combat, or where he makes use of a more skilful patriotic propaganda, he wins Ludendorff's admiration. Similarly, he appreciated the German Spartacists, who incidentally made no attempt upon his life, and was far from condemning an enemy who defended himself and acted in accordance with his own interests. Morals exist only in so far as they are defended by arms.

Ludendorff would not have been embarrassed by any question about the purpose underlying such a gloomy universe, where existence was a matter of battle, murder, and sudden death. Here again, he was subject to no doubt. He was convinced that in some way it was the destiny of Germany to obtain a maximum of power, although it would be incorrect to see any similarity between this notion and the Nationalism which played so important a part during the rise of the middle classes and culminated in the nationalistic states of Europe.

Ludendorff's nationalism had for its object, not the German nation as it had evolved in the course of history, but a non-existent and abstract Germany which still remained to be created. For Ludendorff, everything having an affinity with the history of European thought or the least trace of community with its culture and civilisation was suspicious, harmful, and to be destroyed. His whole attitude resembled that old-fashioned Teutonism which on various occasions after the wars of liberation darkened the German horizon, and which has been preserved by patriotic historians in a bowdlerised form, adorned with pictures of Father Jahn and the student movements of an earlier generation. Such relapses into the cult of a pre-Christian morality invariably exert a fascination whenever certain classes within the nation begin to doubt the blessings of civilisation.

This discontent with civilisation is described by Freud as due to the impulse towards aggression. He considers that in normal circumstances it is possible to preserve the fiction of a common culture among a comparatively large community by providing a kind of lightning conductor for this impulse. However, the "narcissism of minor differences", as Freud calls the indulgence of the aggressive impulse within one cultural community, does not suffice in times of grave crisis. The Jewish people, scattered as they are throughout the world have, in the words of Freud "rendered considerable services to the civilisation of the respective nations among which they have dwelt " in that they became the recipients and victims of this aggressive impulse; and even to this day the Jews are the scapegoats of all those who interpret every defeat of brute force as the malevolence of some hidden and wicked power. In discovering the activities of this Jewish bogey, Ludendorff has shown powers of imagination of which normally he is totally devoid. His anti-Semitic sentiment, however, had not the quality which Freud ascribes to it when he says that the function of this sentiment is to protect a Christian cultural community by concentrating the latter's aggression upon itself. His hatred was not strong enough to prevent him from denying the existence of the cultural community; once engaged upon the search after the bogey which had destroyed his sense of security, he declined to be satisfied with the "narcissism of minor differences". Continuing on his way, he begins to suspect the devil behind every mask, loses every standard of judgment, sees in harmless gatherings like those of modern freemasonry a world-organisation of fantastic conspirators, discovers in the Pope the arch-enemy of Teutonism, and ends his wild career in the worship of Wotan.

Willy Haas has written an interesting study in which he

LUDENDORFF IN 1929

attempts to explain this tragic aberration as a kind of religious mania. In his opinion a complete diagnosis of Ludendorff's disease is impossible except in terms of the experiences of theological discipline. This explanation does not fully meet the facts. A religious enthusiast, in whatever way his mania manifests itself, bears plain traces of his origin; but Ludendorff has nothing to do with the fourth dimension. Ludendorff will remain inexplicable, as will many of the manifestations of the secularised variant of Prussian theology, so long as it is not grasped that the heaven of its votaries ends at Potsdam. Ludendorff's deity is an incredibly meagre being. He has created the world so that his creatures—none other than Ludendorff's Germans—might reach perfection. To this end they must wage wars. How wars are to be waged is known to the Prussian General Staff.

What is the origin of this conviction that it must be so and not otherwise? It is due to the profound desire to obey in all matters, acquired in the course of centuries of drill and discipline. The last of a long series of ancestors, each one of whom had grown up in habits of obedience, Ludendorff was incapable of imagining a world where orders were not supreme. The classical attitude, with its scepticism in the face of every question divine, human, or intellectual, eluded him as much as the independent isolation of a free intellect. "Man must require something which he can obey unquestioningly"; such is the foundation on which the security and the good conscience of German subjects repose.

Nietzsche, who spent so much thought upon the theory that a sense of obedience is the source of German ethics, has set it down that in his opinion this sense is a natural property of the Germans. To them everything that fetters the powers of the intellect and unfetters the passions is a source of danger; accordingly the "love of believing and obeying" is more powerful in the Germans than in most races. We know, however, that the German love of obedience has a simpler origin and one historically less distant; it is the product of German history. German ethics are affected by the historical fact that the Germans sublimated compulsory obedience into the ethical postulate for obedience.

T

Thus it is not a natural but an acquired characteristic. It is not the property of the ruling castes, but the virtue of the subject classes. Nietzsche goes on to explain that German men of genius represent so many isolated cases where the impulse towards obedience is broken; in cases of extremity, in moments of courage, when a man sets his teeth, begins to doubt everything, and is faced by the dreadful impossibility of obeying and the necessity of being his own master—in such moments the German rises above his customary ethics and becomes capable of the highest. In evolving this explanation Nietzsche, the clergyman's son, no doubt had his own experiences in mind.

A state of despair and universal doubt is not, however, a guarantee that the person so afflicted shall rise to any unheard of heights. It is possible to escape from the dungeons of German moral theory, but such an escape is not the inevitable prelude of the birth of genius. A touch of genius must be in existence beforehand; where it exists, it breaks through the bonds of education and, "faced by the dreadful impossibility of obeying", its owner does in fact become capable of the highest. Such, however, is not the case where the process takes place within a petty personality, whose feeling of security is based upon concrete knowledge and on professional fanaticism, in other words, in a German expert. Assailed by doubts, such a person will not be compelled to revise his knowledge and views and to wrestle with God; being of the earth earthly, he will look for the cause of his private catastrophe in earthly factors. He adopts a more moral tone than he had before, accuses others, and looks for the source of his trouble in a genuine conspiracy of fiends. The religious maniac does not conduct polemics against the fiend by whom he is possessed, but, on the contrary, derives a certain satisfaction from the illusion that he is capable of perceiving the depths of existence.

Ludendorff's tragedy is of a deeper nature. In all the course of his errors he does not succeed in escaping from the three-dimensional world. Whose fault is it that his conviction failed when the great test came? Who was it who spoilt the master's masterpiece? Ludendorff's was the most common-

THE END

place of aberrations, and it is indifferent whether he puts the blame on Jews or Freemasons, Pope or Jesuits.

Nevertheless the case of Ludendorff remains interesting because it shows that there are certain fundamental desires which are ineradicable and not amenable to reason. He suffered shipwreck as the apostle of his own gospel; but the reason for his catastrophe was not that his former followers had understood and seen through the fundamental error vitiating his self-destructive course; he became lonely because he was consistent in this last of his illusions, and because the disciples of the national resurrection preached by him declined to follow him as far as Wotan. His fundamental errors, however, are alive in millions; they have had the strength to give rise to political parties and to armed forces, and the day may yet come when they will be a real power. Power cannot be killed by ridicule, for, in Napoleon's words, power is never ridiculous. Whether it rests upon the errors of individuals or upon collective illusions it becomes dangerous as soon as bayonets are at its disposal.

There are to-day millions of people whose humdrum world is brightened by Ludendorff's speculations and by his teleology with all its consequences. Political events have not contributed much to correct this picture. A complaisant pacificism as preached by Count Coudenhove-Kalergi has done little to spread more humane ways of thought; the optimism of Liberal Democrats, in the ears of those who do not benefit under the present order, sounds merely like an expression of satisfaction on the part of those who do benefit by a system which is far from being beyond challenge; finally, there is an enormous army of those who have lost caste and moral fibre through the war, and who have no material interest in the preservation of the present system.

There can be no doubt that men are becoming more ready to see in force the only means of bringing about a change. This powerful craving is deaf to the tender notes of humanitarianism. In face of such cravings one can only repeat the lesson of the Great War, which teaches that force without intelligence is doomed to defeat. To organise, to intensify, and to develop force, those valuable second-rate faculties suffice which are all too common in a nation accustomed to obedience and infested by experts. Victory requires more than this; it demands the use and esteem of the highest of all powers, namely, of human intelligence. Defeat following great victories leads to a variety of superstitions, the most dangerous of which is the fantastic belief that evil is always lurking in ambush. Such a belief is an attempt to escape from the painful truth that conflict does not know good or evil but only degrees of cleverness. Prussia and Germany owe their rise during the last century to the intelligence which guided the sword.

It is the tragedy of a great commander like Ludendorff that he was a sword which there was no intelligence fit to wield.

BIBLIOGRAPHY

FEW points remain unexplained in the history of the war as carried on on the German side. The notion that time must elapse before an opinion can be formed about historical events is the legacy of the period when historians were dependent on the goodwill and generosity of the Government which held the keys of the archives. German strategy during the Great War, from the political as well as the military point of view, has been subjected to exhaustive studies. The work produced by the Reichstag Committee of Investigation, under the title of *Die Ursachen des deutschen Zusammenbruchs im Jahre 1918* (Deutsche Verlagsgesellschaft für Politik und Geschichte, Berlin), has collected all the relevant events with exemplary accuracy. There is further available a large collection of memoirs containing an account of the activities and judgments of those who played a historic part during this period.

In the present work use has been made of Ludendorff's own works as well as of the official archives, the memoirs of Bethmann Hollweg, Kühlmann, Michaelis, Helfferich, Hindenburg, Falkenhayn, General Hoffmann, Colonel Bauer, the younger Hertling, Colonel Nicolai, Lieutenant-Colonel Niemann, Friedrich von Payer, Moltke, the Chief of General Staff, and, above all, of the publications of the Committee of Investigation. The account of the events of 1918 is based on the report published by Colonel Schwertfeger and on General von Kuhl's exhaustive analysis of the various offensives; reference has also been made to Ludendorff's papers, to various documents of the General Staff, and ex-enemy sources like the works of Churchill, Foch, Clemenceau, Pierrefeu, Repington, Page, and others.

Frequent mention has been made of Hans von Hentig's unique work, *Psychologische Strategie des Grossen Krieges* (Heidelberg, 1927, Carl Winter). The part played by the Reichstag during the war is dealt with exhaustively in Professor Bredt's report (vol. viii. of *Ursachen des deutschen Zusammenbruchs im Jahre 1918*). A good insight into the connection between war and politics is given by Dr. Arthur Rosenberg's *Die Entstehung der Deutschen Republik* (Rowohlt, Berlin), which we have frequently quoted. Hans Delbrück's charges against Ludendorff have been drawn from his report (produced concurrently with von Kuhl) and from other of his works. The part which attempts to explain the classical theory of the Prussian military academy is based on a study of the voluminous military literature of Germany, especially on the works of Moltke and Schlieffen.

INDEX

Alberich Line, 96, 97
Arnim, General Sixt von, 204
Artamanov, General, 11
Arz, General, 64
Asquith, Mr. H. H., 67

Bauer, Colonel, 42, 56, 59, 61, 92, 94, 95, 107, 108
Below, General Fritz von, 39
—— General Otto von, 43, 62, 179, 186
Benedek, General, 9
Berthelon, General, 65
Bernstorff, Count, 67, 112
Beseler, General von, 72, 76
Bethmann Hollweg, 52, 56, 67, 71, 72, 73, 74, 75, 76, 80, 81, 82, 83, 84, 86, 87, 88, 89, 105, 107, 108, 109, 110, 112, 113, 114, 125, 126, 127, 157, 158, 160, 163
Bismarck, 36, 47, 48, 49, 50, 58, 69, 70, 75, 77, 88, 89, 91, 105, 111, 145, 149, 152, 158, 212
Blücher, 28
Bockelberg, Major von, 42, 60
Boehn, General, 261
Boraston, Colonel, 134
Borchardt, Rudolf, 101
Bredt, Dr., 75, 111, 113, 138, 144, 220
Brest-Litovsk, peace of, 145 sqq., 161
Brunhilde Line, 242, 261
Brussilov, General, 53, 62
Bülow, General von, 3, 11
—— Prince, 112, 142
Burian, Baron, 67, 73, 244
Bussche, Freiherr von der, 57, 255

Cannae, Battle of, 7, 10, 21, 40, 46, 174
Capelle, Admiral von, 81, 139

Chemin des Dames, attacks on, 100
Churchill, Mr. W. S., 13, 17, 62, 65, 66, 79, 80, 81, 82, 93, 94, 120, 122, 123, 125, 135, 136, 162, 163, 165, 167, 185, 192, 198, 204, 206, 207, 209, 225, 226, 229, 241
Clausewitz, 5, 24, 49, 50, 51, 52, 93, 168, 174, 203, 208, 221
Clemenceau, G., 67, 119, 120, 146, 168, 176, 193, 197, 198
Conrad, General, 29, 31, 35, 40, 51, 53, 54, 63, 73
Cramon, General, 244
Czernin, Count, 45, 104, 127, 131, 146, 149, 160

Delbrück, Hans, 7, 10, 23, 40, 173, 194, 195, 198, 203, 221, 229, 263
Deutelmoser, Herr, 218

Ebert, Herr, 262
Eichhorn, General, 43, 44, 161
Emmich, General von, 1, 2, 3
Erzberger, Herr, 104, 105, 106, 107, 127, 128, 130, 131, 159, 252, 261, 266
Ewert, General, 53

Falkenhausen, General von, 129
Falkenhayn, General von, 29, 30, 32, 34, 39, 40, 41, 42, 43, 45, 47, 51, 52, 53, 54, 55, 56, 59, 63, 64, 65, 75, 80, 171, 175
Favre, Jules, 91
Foch, Marshal, 28, 120, 193, 194, 206, 209, 227, 229, 230, 240, 261
Francis Joseph, 31
François, General von, 13, 14, 15, 16, 17, 18
Fuller, Colonel, 125

Gallwitz, General, 42, 43, 45, 180, 241
George, D. Lloyd, 67, 96, 119, 120, 122, 146, 165, 168, 176, 177, 193, 197, 198
Georgette plan, 200
Gneisenau, 6, 28
Gough, General, 187
Gouraud, General, 226
Groener, General, 47
Grünau, Freiherr von, 211, 254

Haeften, Colonel von, 168, 211, 212, 216, 217, 219, 259, 267
Haig, Field-Marshal, 98, 193, 204, 230, 264
Hartmann, Cardinal, 129
Hausmann, C., 220, 261, 262, 266
Helfferich, Dr., 109, 129, 141, 218
Hentig, Hans von, 20, 32, 45, 49, 52, 99, 102, 209, 232
Hermann Line, 242, 261
Hertling, Count, 110, 112, 121, 142, 143, 144, 145, 149, 150, 155, 156, 157, 159, 161, 168, 197, 201, 215, 216, 217, 220, 221, 233, 234, 235, 237, 239, 243, 244, 245, 246, 251
Heye, Colonel, 267
Hindenburg, General von, 4, 5, 15, 29, 30, 35, 39, 42, 54, 55, 58, 59, 60, 69, 74, 82, 92, 109, 110, 112, 113, 116, 129, 130, 131, 148, 149, 151, 152, 153, 154, 155, 156, 157, 167, 171, 184, 196, 199, 200, 201, 219, 233, 234, 247, 249, 250, 257, 258, 259, 265, 267
Hintze, Herr von, 223, 233, 234, 235, 236, 237, 239, 244, 245, 246, 248, 249, 250, 251, 252, 259
Hitler, Herr, 270
Hoffmann, General, 4, 11, 14, 16, 17, 34, 41, 42, 44, 146, 147, 148, 159, 160, 201
Hohenlohe, Prince, 243
Holtzendorff, Admiral von, 103
Hunding Line, 242, 261
Hutier, General von, 179, 187

Joffé, M., 147
Joffre, Marshal, 95, 96, 97

Kapp, Herr, 139, 270

Kerensky, 138
Kuhl, General, 171, 172, 180, 188, 192, 194, 199, 226, 228, 264, 265
Kühlmann, Dr. von, 126, 127, 128, 129, 139, 145, 146, 147, 149, 152, 154, 158, 159, 160, 161, 211, 212, 217, 218, 219, 220, 222, 223

Lauenstein, General von, 2
Lenin, 103
Lerchenfeld, Count, 142
Linsingen, General von, 35, 37, 161
Lossberg, General von, 207
Lyncker, General von, 55, 56, 57, 110, 112, 113, 148
LUDENDORFF, at Liége, 1, 4
  early life, 5, 6
  Battle of Masurian Lakes, 9 *sqq.*
  Battle of Tannenberg, 12, 13, 14
  attached to infantry, 27
  on Eastern Front, 30 *sqq.*
  called to Pless, 54
  moves to Brest-Litovsk, 55
  appointed to Supreme Command, 57
  visits Western Front, 60
  organises Roumanian campaign, 62
  and Wilson's suggested peace mediations, 68
  and the Polish question, 72 *sqq.*
  and submarine warfare, 79 *sqq.*
  rise to dictatorship, 84 *sqq.*
  sets up War Department, 94
  withdraws to Hindenburg Line, 96
  and the Peace Resolution, 117
  Moldavian campaign, 123
  Italian campaign, 124
  and Belgian question, 129
  deals with Battle of Cambrai, 134
  and Russian Revolution, 137
  at Council of Kreuznach, 145
  threatens resignation, 148
  interferes in question of Eastern frontiers, 152
  submits on frontier question, 156
  prepares for battle in West, 1918, 167 *sqq.*
  arrives at Avesnes for battle, 184

INDEX 281

LUDENDORFF (*continued*)
launches attack, 185
delivers April offensive, 204
delivers May offensive, 214
adopts von Haeften's peace plan, 217
delivers Third Battle of Marne, 224
comment on 8th August, 1918, 231
at Avesnes, 14th August 1918, 233 *sqq.*
at the end of military resources, 247
meets von Hintze at Spa, 249
urges armistice offer, 255
resigns, 267
general criticism, 269 *sqq.*

Mackensen, General, 11, 33, 43, 63, 65
Mars plan, 172
Marschall, Freiherr von, 234
Marwitz, General von der, 187
Masurian Lakes, 9, 12
Marx, Karl, 101
Maunaury, General, 22
Max, Prince, of Baden, 253, 254, 255, 257, 258, 260, 265, 266
Michael plan, 172, 173, 175, 179, 199, 200
Michaelis, Dr., 112, 113, 114, 116, 117, 118, 119, 121, 126, 127, 128, 129, 130, 133, 134, 139, 140, 141, 142, 159, 168, 197
Militia system, 24, 25
Milner, Lord, 193
Moltke, General von, jnr., 3, 4, 8, 19, 21, 22, 29, 59
Moltke, General von, snr., 6, 8, 17, 18, 19, 20, 26, 27, 45, 46, 48, 49, 50, 51, 89, 90, 91, 121, 149, 152, 174, 199, 212
Moser, General von, 173, 174
Mühlmann, General, 14

Nicolaievitch, Grand Duke Nicolai, 30, 33, 35, 37, 39, 42, 44
Niemann, Lieut.-Colonel, 233, 249
Nietzsche, F., 115
Nivelle, General, 96, 97, 98, 100, 120, 177
Northcliffe, Lord, 67

Pacelli, Monsignor (now Cardinal) 125, 126, 127, 130, 163
Payer, Friedrich von, 108, 114, 128, 144, 161, 217, 239, 243, 246, 252, 253, 256, 257, 261, 266
Pershing, General, 202, 230
Pétain, General, 100, 193, 227, 230
Plessen, General, 56, 57, 112, 113, 234
Pierrefeu, 10
Prittwitz, General von, 4
Protopopoff, M., 73

Quast, General von, 204

Radowitz, Herr von, 217
Rathenau, W., 261
Rawlinson, General, 230
Rennenkampf, General, 4, 9, 10, 11, 12, 13, 16
Richthofen, Freiherr von, 72, 106
Rizoff, M., 109
Robertson, Sir W., 165
Roedern, Count, 251, 258
Rosenberg, Dr. A., 76, 77, 82, 114, 220, 237, 239, 248, 251, 253

St. George Plan I., 171, 199, 200
—— II., 172, 199, 200
Samsanov, General, 4, 8, 9, 10, 14
Scheer, Admiral, 266
Scheidemann, General, 33
—— Herr, 159, 162, 163, 262, 266
Scheuch, General, 266
Schlieffen, Alfred von, 5, 6, 7, 8, 20, 21, 22, 23, 24, 26, 27, 34, 40, 44, 45, 46, 47, 52, 174, 195, 199, 208, 209
Schmaus, Lieut., 184
Scholtz, General, 42, 43, 45
Schulenburg, General von der, 171, 172, 180, 188, 241
Schulz, Hugo, 24
Schwertfeger, Colonel, 47, 217, 256
Seeckt, General von, 54
Solf, Dr., 257, 261, 262, 266
Spengler, Oswald, 101
Stein, General von, 4, 25
Steinmetz, General von, 18
Stendhal, 3, 9
Stinnes, Hugo, 73
Stoss, General, 91

Stresemann, G., 70, 106, 107, 108, 127, 128, 140, 252, 255, 256
Stumm, Herr von, 246
Stürmer, M., 75
Südekum, Herr, 116

Tannenberg, Battle of, 10, 11, 13, 14, 16, 175
Tappen, Colonel von, 4, 11, 12, 56, 59
Tirpitz, Admiral von, 80, 81, 139
Trimborn, Herr, 261, 266
Trotsky, L., 160, 161

Units of German Army:
  1st Army, 181, 214, 224, 225, 226
  2nd Army, 3, 180, 181, 182, 187, 188, 189, 190
  3rd Army, 181, 224, 225, 226
  4th Army, 181, 188, 204, 205, 206, 207
  6th Army, 181, 188, 190, 204
  7th Army, 180, 181, 190, 214, 224
  8th Army, 4, 14, 30, 33, 37
  9th Army, 30, 33, 38, 41, 63
  10th Army, 43
  17th Army, 180, 181, 182, 186, 187, 188, 189, 190, 198
  18th Army, 180, 181, 182, 187, 188, 189, 190, 194, 215
  1st Army Corps, 13, 15
  17th Army Corps, 15
  20th Army Corps, 15, 16
  Bavarian Alpine Corps, 63, 64
  1st Division, 16
  2nd Division, 16
  41st Division, 16
  1st Reserve Corps, 15

—— of British Army:
  3rd Army, 185
  4th Army, 220
  5th Army, 185, 187
  9th Division, 185
—— of Russian Army:
  1st Army Corps, 15, 17

Valentini, Herr von, 112, 157

Waldersee, Count, 4, 19, 20
Westarp, Count, 219, 220, 255, 256
Wetzell, Colonel, 59, 124, 125, 171, 195, 203
William I., 25, 26, 45, 46, 88, 90, 152, 212
William II., 19, 25, 28, 31, 35, 39, 42, 45, 55, 56, 58, 67, 68, 72, 77, 80, 82, 83, 102, 105, 108, 112, 113, 118, 119, 125, 126, 129, 130, 131, 140, 142, 145, 147, 150, 151, 152, 153, 154, 156, 157, 158, 159, 167, 169, 170, 196, 197, 203, 221, 222, 223, 233, 235, 245, 249, 250, 251, 255, 257, 266, 267
Wilson, President, 67, 92, 162, 163, 168, 202, 245, 246, 249, 250, 258, 259, 260, 261, 262, 265, 266
—— Sir H., 165, 193
Wotan Line, 243
Woyrsch, General, 179
Wright, Captain, 179, 192

Yorck, General, 28

Zimmermann, Dr., 73, 92

THE END